Masoud Aryanpour

Ab-Initio Modelling of Electrochemistry

Masoud Aryanpour

Ab-Initio Modelling of Electrochemistry

Application to Proton-Exchange-Membrane Fuel Cells

VDM Verlag Dr. Müller

Imprint

Bibliographic information by the German National Library: The German National Library lists this publication at the German National Bibliography; detailed bibliographic information is available on the Internet at http://dnb.d-nb.de.

Cover image: www.purestockx.com

Publisher:
VDM Verlag Dr. Müller Aktiengesellschaft & Co. KG , Dudweiler Landstr. 125 a, 66123 Saarbrücken, Germany,
Phone +49 681 9100-698, Fax +49 681 9100-988,
Email: info@vdm-verlag.de

Zugl.: Stanford, Stanford University, Diss., 2007

Produced in USA and UK by:
Lightning Source Inc., La Vergne, Tennessee, USA
Lightning Source UK Ltd., Milton Keynes, UK
BookSurge LLC, 5341 Dorchester Road, Suite 16, North Charleston, SC 29418, USA

ISBN: 978-3-639-03857-6

To my parents, *Mohammad* and *Shokat*

Preface

Generating power has been and will remain a basic need for human life. Instead of burning hydrocarbons, fuel cells are presented to be an alternative solution for the power generation problem by having the least amount of harmful effects. In current fuel cell models, chemical processes are coupled to transport and flow processes through simplified models, which neglect the details of real chemistry and thus fail to predict the correct trends of surface coverage in fuel cells.

Multi-scale models for PEM fuel cells can be developed, in which not only transport and flow, but also the chemistry is simulated by the best available computational tools, here the Dynamic Monte Carlo (DMC) methodology. DMC simulations can predict the state of the catalyst surface quite accurately if they are provided by accurate rate constants for elementary reactions that are randomly chosen from a chemical mechanism based on their relative probabilities.

Ab-initio calculations have been used here to provide the necessary kinetic rate data for performing DMC simulations of oxygen reduction reaction (ORR) mechanism in PEM fuel cells. An efficient search algorithm is developed to locate the transition states of electron transfer reactions for large systems, for which such calculations were not feasible before. A set of consistent energetics are presented for the elementary electrochemical reactions in the ORR mechanism. Due to their prominent role, interaction of adsorbed O and OH on these reactions are investigated too. Using these data, the DMC results for the water discharge mechanism, the major subset of the ORR mechanism, is in very good agreement with experimental measurements. Whenever possible, important concepts have been discussed to provide further insights and help in future applications of quantum calculations in electrochemistry.

iv

Nomenclature

Greek Letters

α, β	spin functions in quantum chemistry (QC), or symmetry factors in electrochemistry (EC)
δ	Kronecker delta function
ϵ_o	permittivity of vacuum
ε	energy attributed to an electron when occupying an orbital
ζ	coefficient of length in the exponents of orbital functions
η	overpotential
λ	reorganization energy
μ	Lagrange multiplier
ρ	electronic probability density
σ	spin variable
ϕ	molecular space-orbital in QC, or electrode potential in EC
φ	molecular spin-orbital in QC, or activation energy in EC
χ	atomic spin-orbital
ψ	chemical potential, stands equally for IP or EA
Ψ	wave function
ω	vibrational frequency
Ω	space or subspace

Latin Letters

A	pre-exponential factor in the Boltzmann Arrhenius form for rate constants
b	temperature exponent in the Boltzmann Arrhenius form for rate constants
e	electronic charge
$\mathbf{e}$	unit vector
E	energy of a molecular system
E_e	work function of an electrode
F	Faraday's constant
$\mathbf{F}$	Fock matrix
$\mathbf{G}$	inverse of the mass matrix
h	Planck's constant in J.s
h	one-electron integral in the Hartree-Fock approximation
$\mathbf{H}$	Hamiltonian, or Hamiltonian matrix
i	dummy index, or electrical current current, or $\sqrt{-1}$
j	an index, electrical current density
J	classical Coulomb's potential energy in DFT formulation
J, K	two-electron integrals in the Hartree-Fock approximation
k	reaction coefficient
$\mathbf{k}$	wave vector
k_B	Boltzmann constant $(1.3806503 \times 10^{-23}$ m^{2}kg s^{-2} K$^{-1})$
$\mathbf{K}$	stiffness matrix
$\mathcal{L}$	Lagrange function
m	mass
m_e	mass of an electron
$\mathbf{M}$	mass matrix
N	number of degrees of freedom, or number of objects
P	probability
$\mathbf{q}$	vector of normal coordinates

Q	vector of canonical variables
R	a position vector, especially for a nucleus
R	universal gas constant
r	a position vector, especially for an electron
s	spin
S	overlap matrix
t	time
T	absolute temperature in Kelvin, or kinetic energy
U	internal energy in QC, electrode potential in EC
x, y, z	coordinates in Cartesian system
x	coordinates vector, or vector of variables
X	vector of structural variables
Z	atomic number

Operators

$\Delta \cdot$	change in variable $\cdot$
$d\cdot$	differential of variable $\cdot$
$\cdot^*$	complex conjugate of $\cdot$
$\hat{\cdot}$	quantum mechanical operator corresponding to $\cdot$
$\tilde{\cdot}$	candidate solution for quantity $\cdot$
$\delta \cdot$	variation of $\cdot$

Abbreviations

AFC	Alkaline Fuel Cell
AO	Atomic Orbital
BO	Born-Oppenheimer
CGO	Contracted-Gaussian-Orbital
DFT	Density Functional Theory
DMC	Dynamic Monte-Carlo
DMFC	Direct Methanol Fuel Cell
DOF	Degrees Of Freedom
eV	Electron Volts
EA	Electron affinity
EC	Electrochemistry
ECP	Effective Core Potentials
ET	Electron Transfer
GGA	Generalized Gradient Approximation
GTO	Gaussian-Type-Orbital
HF	Hartree-Fock
IP	Ionization potential
LCAO	Linear Combination of Atomic Orbitals
LDA	Local Density Approximation
LRC	Local Reaction Center
LSD	Local Spin Density
MCFC	Molten Carbonates Fuel Cell
MD	Molecular Dynamics

MO	Molecular Orbital
ORR	Oxygen Reduction Reaction
PAC	Partial Atomic Charges
PAFC	Phosphoric Acid Fuel Cell
PEM	Proton Exchange Membrane
PEMFC	Proton-Exchange Membrane Fuel Cell
PES	Potential Energy Surface
QC	Quantum Chemistry
RDS	Rate Determining Step
RHF	Restricted Hartree-Fock
SCF	Self-Consistent Field
SHE	Standard Hydrogen Electrode
SOFC	Solid Oxide Fuel Cell
STO	Slater-Type-Orbital
TS	Transition State
TST	Transition State Theory
UHF	Unrestricted Hartree-Fock
ψDAE	ψ-Dependent Activation Energies

Quantum Theories

B3LYP	DFT hybrid functional based on the three-parameter formula of Becke and the correlation functional of Lee-Yang-Parr
CCSD	Coupled Cluster method including Singles and Double
CI	Configuration Interaction
MP	Møller-Plesset perturbation method
QCISD	Quadratic Configuration Interaction including Singles and Doubles

Basis Sets

aug-cc-pVnZ	cc-pVnZ basis set with diffuse functions
cc-pVnZ	Correlation-Consistent Polarized n-Valence ζ basis sets
LANL2DZ	Los Alamos National Laboratory double-ζ basis set
6-31G	basis set in which every core orbital uses 6 primitive Gaussian functions, and each valence orbital is a double-ζ that has 3 primitives for the first and 1 for the second function
6-31G*	6-31G basis set in which d functions have been added to the p functions
6-31G**	6-31G* basis set in which p functions have been added to s functions of H and He atoms
6-31+G	6-31G basis set with one additional s and one additional set of p functions
6-31++G	6-31+G basis set with a diffuse s function added to the atoms on the first row

Contents

Preface   iv

Nomenclature   v

Abbreviations   viii

1   Introduction   1
    1.1   Fuel Cells as Power Sources   1
    1.2   Proton Exchange Membrane Fuel Cells   6
    1.3   Multi-Scale Modeling of PEM Fuel Cells   9
    1.4   Dynamic Monte-Carlo Method and Kinetics   11
    1.5   General Form of the Rate Constants   13
    1.6   Rate Constants from the Literature   15
    1.7   Rate Constants and Quantum Simulations   15

2   Computational Quantum Chemistry   20
    2.1   Schrödinger Equation   21
    2.2   Born-Oppenheimer Approximation   24
    2.3   Molecular Orbital (Wave Function) Models   28
       2.3.1   Slater Determinants   28
       2.3.2   Variation Principle   31
       2.3.3   Hartree-Fock Approximation   34
       2.3.4   Roothaan Equations   38
    2.4   Density Functional Theory   40

	2.4.1	Hohenberg-Kohn Theorems	41
	2.4.2	Kohn-Sham Equations	45
	2.4.3	Exchange-Correlation Functionals	48
2.5	Basis Sets		52
2.6	Comparison of Performance and Accuracy		62
2.7	Summary		66

3 Electrochemistry in PEM Fuel Cells — **67**

3.1	Literature Review	68
3.2	Macroscopic View in Electron Transfer	72
3.3	Microscopic View: Marcus Theory	79
3.4	Summary	84

4 Transition States of ET Reactions — **85**

4.1	Local Reaction Center Theory	86
4.2	CICV Method	88
4.3	Principle of Microscopic Reversibility	94
4.4	Convergence of the CICV Method	97
4.5	Summary	99

5 Computational Investigation of LRC Theory — **101**

5.1	Effect of Madelung Charge	101
5.2	Accuracy of Ionization Potentials	109
5.3	Size Effect of Pt Cluster	111
5.4	Summary	114

6 Activation Energies of ET Reactions — **116**

6.1	Consistent Computational Scheme	117
6.2	$OH_{(ads)} \rightleftharpoons H_2O$	121
6.3	$O_{(ads)} \rightleftharpoons OH_{(ads)}$ with and without OH-Interaction	124
6.4	$O_{(ads)} + H^+ + e^- \rightleftharpoons OH_{(ads)}$ with O-Interaction	142
6.5	$O_{2(ads)} \rightleftharpoons O_2H_{(ads)}$	151

6.6 Thermodynamics of Interaction Effects 156

6.7 Application to Water Discharge 160

6.8 Summary 163

7 Prefactors of ET Reactions   **164**

7.1 Theoretical Background 165

    7.1.1 Modal Analysis and Pre-Exponential Factors 166

    7.1.2 Mass Matrix in Terms of Internal Coordinates 168

7.2 Constraints in Mass Matrix Calculations 173

7.3 Normal Mode Transformation 175

7.4 Estimation of Pre-Exponential Factors 177

A Extensions to LRC Theory   **180**

A.1 Exciting Electrode by External Charges 181

A.2 Isolating Reaction Center in ψ Calculations 184

B Equality of ET Prefactors   **191**

B.1 Kinetics of Electron Transfer Reactions 191

B.2 Equilibrium and Standard Conditions 192

B.3 Potential-Dependent Activation Energies 193

B.4 Rate Constants and Equilibrium Current 194

B.5 Nernst Equation 195

B.6 Thermodynamics of Electrochemical Reactions 197

B.7 Rates at Standard & Equilibrium Conditions 199

Bibliography   **201**

List of Tables

1.1   Rate constants for surface reactions, Reference Försth [41]   16

2.1   Scaling behavior of methods versus the number of basis functions N, (Cramer [31])   63

2.2   Mean absolute errors in bond lengths for different methods for the G2 set, (Cramer [31])   64

2.3   Mean absolute errors in enthalpies of the test reaction set reported in kcal/mol,(Cramer [31])   65

2.4   Mean absolute errors in IPs and EAs for test set G2 in eV,(Cramer [31])   66

3.1   Reaction steps constituting the oxygen reduction reaction at cathode electrode of PEM fuel cells.   71

4.1   Degrees of freedom of the precursor in the Pt_2O_2H oxidation reaction (4.18)   98

6.1   Properties of the optimized reactants and products precursors in the oxidation/reduction reactions of Pt-$Pt.OH_2$.   124

6.2   Properties of the optimized reactants and products precursors in the oxidation/reduction reactions of $Pt.OH$-$Pt.OH_2$.   125

6.3   Activation energies at various electrode potentials for the oxidation/reduction reactions of Pt-$Pt.OH_2$.   126

6.4   Activation energies at various electrode potentials for the oxidation/reduction reactions of $Pt.OH$-$Pt.OH_2$.   127

6.5 Reversible potential U_0 for oxidation/reduction reactions of Pt-Pt.OH$_2$ and Pt.OH-Pt.OH$_2$. 128

6.6 Energetics of OH[H$_2$O]$_3$ on Pt$_{10}$ and Pt$_2$ clusters vs. the spin state. These systems are denoted by Pt$_8$-Pt-Pt.OH[H$_2$O]$_3$ and Pt-Pt.OH[H$_2$O]$_3$, and shown in Figure 6.8 and Figure 6.9, respectively. 132

6.7 Averaged surface angles for Pt$_8$-Pt-Pt.OH[H$_2$O]$_3$ precursor obtained from the ground spin states (see Table 6.6). 133

6.8 Properties of the ground state for the Pt-Pt.OH[H$_2$O]$_3$ precursor (see Figure 6.9). 134

6.9 Activation energies for Pt-Pt.O + H$^+$ + e$^-$ $\rightleftharpoons$ Pt-Pt.OH 135

6.10 Energetics of OH-OH[H$_2$O]$_3$ on a Pt$_3$ cluster vs. the spin state. This system is denoted by Pt.OH-Pt.OH[H$_2$O]$_3$-Pt and shown in Figure 6.11. 137

6.11 Averaged surface angles for Pt.OH-Pt.OH[H$_2$O]$_3$-Pt precursor obtained from the ground spin states (see Table 6.10). 138

6.12 Energetics of OH-OH[H$_2$O]$_3$ on a Pt$_2$ cluster vs. the spin state. This system is denoted by Pt.OH-Pt.OH[H$_2$O]$_3$ and shown in Figure 6.12. 138

6.13 Properties of the ground state for the Pt.OH-Pt.OH[H$_2$O]$_3$ precursor (see Figure 6.12). 139

6.14 Activation energies for Pt.OH-Pt.O + H$^+$ + e$^-$ $\rightleftharpoons$ Pt.OH-Pt.OH . . 140

6.15 Energetics of O+OH[H$_2$O]$_3$ on a Pt$_{13}$ cluster vs. the spin state. This system is denoted by Pt$_{11}$-PtO-PtOH[H$_2$O]$_3$ and shown in Figure 6.15. 144

6.16 Averaged surface angles for O+OH[H$_2$O]$_3$ on a Pt$_{13}$ cluster obtained from the ground spin states (see Table 6.15). 144

6.17 Energetics of O+OH[H$_2$O]$_3$ on a Pt$_2$ cluster vs. the spin state. This system is denoted by PtO-PtOH[H$_2$O]$_3$, and is shown in Figure 6.16. 146

6.18 Properties of the ground state for the O$_{(ads)}$+OH[H$_2$O]$_{3,(ads)}$ on a Pt$_2$ cluster (see Figure 6.16). 147

6.19 ψ of the ground states for O$_{(ads)}$ + H$^+$ + e$^-$ $\rightleftharpoons$ OH$_{(ads)}$ with and without interactions. 147

6.20 Activation energies for O$_{(ads)}$+OH[H$_2$O]$_{3,(ads)}$ on a Pt$_2$. 150

6.21 Energetics of Pt.O-Pt.OH[H$_2$O]$_3$ vs. the spin state. 153

6.22 Averaged surface angles for Pt.O-Pt.OH[H$_2$O]$_3$ precursor obtained from
      the ground spin states (see Table 6.21, and Figure 6.19). 154

6.23 Properties of the ground state for Pt.O-Pt.OH[H$_2$O]$_3$ (see Figure 6.19). 154

6.24 Activation energies for Pt.O-Pt.O + H$^+$ + e$^-$ $\rightleftharpoons$ Pt.O-Pt.OH 155

7.1 Normal mode analysis of Pt-PtOH$_2$[H$_2$O]$_3$ corresponding to reaction (6.1)178

7.2 Normal mode analysis of Pt-PtOH[H$_2$O]$_3$ corresponding to reaction (6.2)178

7.3 Normal mode analysis of PtO-PtOH[H$_2$O]$_3^\dagger$ corresponding to reaction (6.3)179

7.4 Pre-exponential factors of electron transfer reactions (6.1)-(6.3) . . . 179

A.1 Ground state IP of OH[H$_2$O]$_3$ determined by B3LYP calculations using
     different basis sets. 185

List of Figures

1.1 Basic design of a fuel cell: the electrolyte (membrane) is sandwiched between the anode and the cathode, which in turn are also connected by an outer circuit.   4

1.2 Schematic of a Proton Exchange Membrane (PEM) fuel cell.   7

1.3 Illustrative structure of the cathode of a PEM fuel cell.   8

1.4 Structure of Nafion, the sulfonic ends can sustain water chains, while the remaining parts are hydrophobic.   8

1.5 Schematic of a PEM fuel cell voltage versus current.   10

1.6 Change in the Gibbs free energy during a reaction. The activated complex is the configuration of maximum Gibbs energy along the reaction path that connects the reactants to the products.   14

2.1 Comparison of the radial component of $1s$-STO with a $1s$-GTO . . .   56

2.2 Comparison of the radial component of $1s$-STO with contracted STO-LG, $L = 1, 2, 3$ orbitals.   58

3.1 Normalized current density versus overpotential at room temperature according to the Butler-Volmer equation (3.16)   78

3.2 Potential energy surfaces of redox centers according to the Marcus theory.   81

4.1 Graphical demonstration of the CV algorithm: the CV method locates x_{final}, which corresponds to a slightly different IP (or EA) from the target work function of the electrode ψ_2   91

4.2 Graphical demonstration of the CICV algorithm: the CICV method solves the Lagrange equations such that at each iteration the IP (or EA) of the molecule converges to ψ_2 91

4.3 Reaction center for oxidation of $PtOH_2.(H_2O)_3$ along with five interatomic distances used to optimize the transition states. R4 and R5 distances were changed symmetrically for both of the two hydrogen-bonded water molecules 93

4.4 Activation energy at different potentials for oxidation of $PtOH_2.(H_2O)_3$ from the CICV method compared with the CV method 94

4.5 The reaction center for Pt_2O_2H oxidation/reduction reactions: Pt_2O_2H attached to three water molecules to account for the effect of the solution 95

4.6 Calculated and derived activation energy curves for Pt_2O_2H reactions after the first series of computations using the CICV method to determine the transition states 96

4.7 Number of Iterations per step required for the oxidation reaction of Pt_2O_2H 99

5.1 Position and magnitude of the Madelung charge that is used in the LRC theory to represent the effect of ions in the solution. 102

5.2 Coordinate system used to plot the change in the IP of PtOH in Figure 5.3. 103

5.3 IP of PtOH on the left as affected by a $+0.5e$ point charge that is placed on the x-y plane of the molecule depicted in Figure 5.2. 104

5.4 Electrostatic interactions between a point charge and a PtOH molecule as computed by the Coulomb formula. 106

5.5 Comparison of the quantum-computed IP with the analytical estimations as a function of the ion-charge distance. 107

5.6 Comparison of the quantum-computed IP with the analytical estimations as a function of the charge magnitude when placed at 10Å away from the ions. 108

5.7 Change in the partial atomic charges between PtOH and PtOH$^+$ in terms of the magnitude of a point charge placed at 10Å away from the ions. 108

5.8 Comparison of the ionization potential of an oxygen atom obtained by different computational methods and basis sets with the experimental measurements. 110

5.9 Ionization potential calculation of Pt$_n$OH molecules for $n = 1, 2, 3$ (left) Pt$_1$OH, (middle) Pt$_2$OH, (right) Pt$_3$OH. 111

5.10 Ionization potential calculation of Pt$_n$OH molecules for $n = 5, 7$ (left) Pt$_5$OH, (right) Pt$_7$OH. 112

5.11 Ionization potential calculation of Pt$_{10}$OH molecule. 112

5.12 IP of two molecules Pt$_n$-OH and Pt$_n$ as a function of the number of platinum atoms n. 113

6.1 Definition of surface angles and their importance for optimization of systems with two Pt atoms. 119

6.2 Relaxation of the surface angle in Pt-Pt.OH$_2$ takes the molecule from the initial structure (a) to the unphysical configuration (b). 119

6.3 Auxiliary Pt atoms (represented by dashed lines) are added to prevent structural distortions. 120

6.4 Optimized structure of Pt-Pt.OH$_2$ (a), and that of Pt.OH-Pt.OH$_2$ (b). 123

6.5 PDAE curves for Pt-Pt.OH$_2$ (a), and those of Pt.OH-Pt.OH$_2$ (b). . . 128

6.6 Activation curves for the oxidation/reduction steps of OH$_{(ads)}$ with and without interaction. 129

6.7 Optimization of Pt-Pt.OH[H$_2$O]$_3$-Pt from the initial structure shown in (a) led to the distortion of the geometry shown in (b). 130

6.8 Optimization of Pt$_8$-Pt-Pt.OH[H$_2$O]$_3$ and averaging the surface angles: (a) initial structure, (b) optimized structure with labels used to define the geometry (see Table 6.7). 131

6.9 Optimization of Pt-Pt.OH[H$_2$O]$_3$ structure using the averaged surface angles: (a) atomic symbols, (b) labels used to define the geometry (see Table 6.8). 133

6.10 Potential-dependent activation energies for Pt-Pt.O + H$^+$ + e$^-$ $\rightleftharpoons$ Pt-Pt.OH. The reversible potential at 1.526 V is consistent with its thermodynamic estimation. 136

6.11 Optimization of Pt.OH-Pt.OH[H$_2$O]$_3$-Pt and averaging the surface angles: (a) molecular structure, (b) the same structure with the labels used to define the geometry (see Table 6.11). 137

6.12 Optimization of Pt.OH-Pt.OH[H$_2$O]$_3$ and averaging the surface angles: (a) molecular structure, (b) the same structure with the labels used to define the geometry (see Table 6.13). 139

6.13 Potential-dependent activation energies for Pt.OH-Pt.O + H$^+$ + e$^-$ $\rightleftharpoons$ Pt.OH-Pt.OH. The reversible potential at 1.74 V is consistent with its thermodynamic estimation. 141

6.14 PDAE curves for the oxidation/reduction steps of O$_{ads}$ with and without interaction. 141

6.15 Optimization of surface angles for reaction O$_{(ads)}$ + O$_{(ads)}$ + H$^+$ + e$^-$ $\rightleftharpoons$ O$_{(ads)}$ + OH$_{(ads)}$: initial structure with atomic symbols, and final structure with labels to define the geometry in Table 6.16. 143

6.16 Structure of the reaction center for O$_{(ads)}$ + O$_{(ads)}$ + H$^+$ + e$^-$ $\rightleftharpoons$ O$_{(ads)}$ − OH$_{(ads)}$: (left) the structure of the molecule with atomic symbols, (right) the molecule with labels to define the geometry in Table 6.17. 145

6.17 Potential-dependent activation energies for O$_{(ads)}$ + O$_{(ads)}$ + H$^+$ + e$^-$ $\rightleftharpoons$ O$_{(ads)}$ + OH$_{(ads)}$. The oxidation curve (oxi.$_{ca.}$) was used to obtain the reduction activation energies denoted by red.$_{deriv.}$ 149

6.18 Comparison of PDAE curves for reaction O$_{(ads)}$ + H$^+$ + e$^-$ $\rightleftharpoons$ OH$_{(ads)}$ with and without the interaction of an O$_{(ads)}$. 151

6.19 Optimization of Pt.O-Pt.OH[H$_2$O]$_3$ structure: (a) atomic symbols, (b) labels used to define the geometry (see Table 6.23) 153

6.20 Potential-dependent activation energies for Pt.O-Pt.O + H$^+$ + e$^-$ $\rightleftharpoons$ Pt.O-Pt.OH. The reversible potential at 0.951 V is consistent with its thermodynamic estimation. 155

6.21 System energy without and with interaction effects as a function of reaction coordinate 157

6.22 Residual of equation (6.17) for the OH interaction with the electrochemical reactions of O$_{(ads)}$ and OH$_{(ads)}$. 161

6.23 Comparison of the coverage of O containing species versus electrode potential for water discharge mechanism [101]. 162

7.1 (left) Illustration of s_{t1} and s_{t2} as an increase in the bond length r_{12} between atoms 1 and 2, and (right) illustration of s_{t1}, s_{t2}, and s_{t3} as an increase in the bond angle ϕ formed between atoms 1, 2, and 3. . . 169

7.2 Illustration of dihedral angle τ between two planes formed by atoms 1, 2, 3 and 2, 3, 4. 170

7.3 (left) Geometrical solution to a 2-fold constraint problem of combining s_1 and s_2, and (right) geometrical solution to a 3-fold constraint problem of combining s_1, s_2, and s_3. 171

A.1 Exciting Pt cluster with a point charge in Pt$_1$OH[H$_2$O]$_3$, (left) molecular structure and the point charge, denoted by X, (right) IP of Pt$_1$ and Pt$_1$OH[H$_2$O]$_3$ as a function of the charge magnitude. 182

A.2 Exciting Pt cluster with point charges in Pt$_{3,1}$OH[H$_2$O]$_3$, (left) molecular structure and the point charges, denoted by X, (right) IP of Pt$_{3,1}$ and Pt$_{3,1}$OH[H$_2$O]$_3$ as a function of the total charge magnitude. . . . 183

A.3 Exciting Pt cluster with point charges in Pt$_{7,3}$OH[H$_2$O]$_3$, (left) molecular structure and the point charges, denoted by X, (right) IP of Pt$_{7,3}$ and Pt$_{7,3}$OH[H$_2$O]$_3$ as a function of the total charge magnitude. . . . 184

A.4 Activation energy for oxidation of OH[H$_2$O]$_3$ molecule as a function of its IP using different Pt clusters and basis sets. All IP's have been shifted to 11.61 eV, which is the ground state IP of OH[H$_2$O]$_3$: B3LYP/6-31G*. 186

A.5 Activation energy for reaction $OH_{(ads)} \rightleftharpoons O_{(ads)} + H^+_{(aq)} + e^-$ using $Pt_{3,1}OH[H_2O]_3$: B3LYP/6-31G*/LANL2MB. 187

A.6 Activation energy for oxidation of $OH[H_2O]_3$ molecule on a $Pt_{3,1}$ cluster as a function of its IP using different number of Pt atoms in the IP calculations. All IP's have been shifted to 11.61 eV, which is the ground state IP of $OH[H_2O]_3$: B3LYP/6-31G*. 189

Chapter 1

Introduction

Generating power has been and will remain a basic need for human life. During the past centuries, burning hydrocarbons to satisfy this need has left negative and serious effects on the earth and its climate. Fuel cells are an alternative solution for the power generation problem, having the least amount of harmful effects. The widespread use of fuel cells requires fundamental improvements in their structure and operation, an intersection that challenges both our current technology and scientific knowledge. Quantum computations can facilitate the modeling of the involved chemistry, and serve as a predictive tool to improve current fuel cells.

1.1   Fuel Cells as Power Sources

Fuel cells are known as the ideal power generators of the future. A fuel cell is basically an energy converter that, similar to a battery, converts chemical energy into electricity. However, a cell functions in a superior way than a battery. While the performance of a typical battery drops over time, a fuel cell can generate almost constant output power as long as it is provided with fuel and oxygen. In hydrogen fuel cells, the chemical reaction is the combination of hydrogen with the oxygen that exists in the air or is supplied as liquid. This combination process involves the transfer of electrons between the two sides of the cell, which produces electricity in the outer circuit (Figure 1.1). In addition to electricity, water and heat constitute the main products of a fuel cell.

Higher efficiency and lower emissions are the major advantages of fuel cells over traditional power systems. Efficiency as the measure of performance represents the fraction of the input power that is transformed into the useful output power. The higher the efficiency of a power generator, the more output energy is accessible in the desired form. In automotive applications, internal combustion engines have efficiencies of about 20 %, compared to 45 % for fuel cells. This means that half of the energy consumed by vehicles can, at least in theory, be saved if fuel cells are used instead of combustion engines. The second feature is that fuel cells produce far fewer pollutant gases, such as CO_2 and NO_x, than combustion-based devices. For now, motor vehicles and large factories are the main sources of air pollution in urban and industrial areas. Yet fuel cells can potentially reduce CO_2 emissions by 25-60 % with respect to natural gas and coal power plants, besides being flexible to build up various scaled power stations [65]. Finally, since there is no moving part, a fuel cell virtually produces no noise, a desired feature in environments such as hospitals and domestic applications.

With these many essential benefits, why are fuel cells not in common use today? The idea of fuel cells is not new but dates back to the work by Sir William Grove in the 19th century. His experiments provided evidence that hydrogen and oxygen are both vital to produce electricity in an electrochemical cell. Despite Grove's observations, the development of fuel cells was not pursued because other sources of power that consumed fossil fuels were more feasible and affordable. Due to cost and technological issues, limited application was the fate of fuel cells for nearly one century. Then, in the 1960s, the US National Aeronautics and Space Administration (NASA) used fuel cells for powering spaceships. That pioneering application brought fuel cells to public awareness as a quality power source, but the technology was still too expensive for terrestrial applications.

Despite all the involved challenges, in recent years, fuel cells have received considerable attention both in industry and in research. The non-stop increase in electrical energy demand, shortage of conventional fuels and environmental issues [27] are motivating governmental investments and technological interests. The US Department of Energy (DOE) aims to reduce the current cost of fuel cells from $4,500 per kilowatt to $400 by the end of this decade. This effort, if successful, will make fuel cells competitive with all other power systems, which currently cost about $800-$1,500 per kilowatt.

Several types of fuel cells (FC) are currently available: alkaline (AFC), proton-exchange membrane (PEMFC), direct methanol (DMFC), solid oxide (SOFC), molten carbonates (MCFC) and phosphoric acid fuel cells (PAFC). Each type, except DMFC, is named after its electrolyte, which is the medium that allows ions to move between the electrodes. The electrolyte itself is the same as the membrane or is contained in the membrane, shown in Figure 1.1. For instance, the electrolyte in PAFCs is liquid phosphoric acid contained in a solid matrix such as silicone carbide. DMFCs are a special kind of PEMFCs in which the fuel, methanol, is directly supplied to the cell without prior conversion to hydrogen. All the other types are fed with pure hydrogen or a gas that is highly rich in hydrogen.

Each type of fuel cell has its own characteristics and features, and thus has its own applications. These characteristics cover different categories: operating conditions (e.g., the operating temperature range), materials and design (e.g., the material of the electrodes) and output parameters (e.g., the output power.) MCFCs and SOFCs work at temperatures higher than 900 K, while PAFCs work at around 400 K. PEMFCs operate at slightly higher than room temperature, at approximately 330 K. AFCs can work at a wider range from very low temperatures up to about 370 K [102]. PEMFCs and PAFCs need platinum-based electrodes, which makes them more expensive than other types. Other fuel cells use noble metal electrodes such as silver and nickel, or their oxides and alloys.

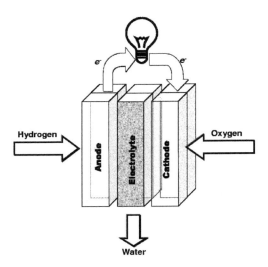

Figure 1.1: Basic design of a fuel cell: the electrolyte (membrane) is sandwiched between the anode and the cathode, which in turn are also connected by an outer circuit.

For the time being, no type of fuel cell is theoretically perfect, nor is any type competitive with commonly used energy sources. Each one has its own limitations and relative disadvantages that make it suitable only for certain applications. Normally, high-temperature fuel cells have to be stationary and corrode faster than other types, although they have higher efficiencies and output power. Low-temperature cells can be portable but are more sensitive to the electrodes material and impurities in the fuel. Finally, all fuel cells are in general much more expensive than conventional power generators. To realize the dream of having a clean power supply through the use of fuel cells, their overall cost must be reduced substantially. In this regard, many challenges in material science, temperature engineering, surface science, control and their related disciplines need to be addressed and overcome.

Among all types of fuel cells, PEM fuel cells have two features that make this type most attractive for public applications. PEM fuel cells work most efficiently in the range of 60 to 80°C, which is lower by a few hundreds of degrees than the working temperature of SOFCs and MCFCs. Operating at low temperatures allows PEMFCs not only to avoid serious technological issues such as corrosion and the need for peripheral devices, but also to weigh less and serve as portable power supplies.

The second feature of PEM fuel cells is their capability to produce a wide range of output power from as low as 50 W up to 250 kW. Although AFCs have a similar output, DMFCs produce at maximum 5 kW, and the rest have a minimum output of 50 to 100 kW [27]. It is worth mentioning that a typical household needs about 5 kW of power. As an actual publicized application, in 1997, the Plug Power company started distributing 7 kW PEM fuel cells for powering homes in New York. The combination of the flexibility in output power with low operating temperature has made PEMFCs an attractive energy source for various applications ranging from car batteries to small portable devices. The rest of this chapter briefly describes PEM fuel cells and the involved chemistry, which is the topic of this book.

1.2   Proton Exchange Membrane Fuel Cells

Figure 1.2 illustrates the structure of a PEM fuel cell. The main components are
the inlet/outlet gas channels, the anode electrode, the electrolyte, and the cathode
electrode. On the catalyst surfaces at the anode side, hydrogen molecules dissociate
and produce H^+ ions according to reaction (1.1).

$$H_2 \rightleftharpoons 2H^+ + 2e^- \qquad\qquad (1.1)$$

On the cathode side, oxygen molecules combine with the H^+ ions coming through
the electrolyte, and the electrons that come from the circuit, thereby forming water:

$$O_2 + 4H^+ + 4e^- \rightleftharpoons 2H_2O \qquad\qquad (1.2)$$

The membrane, or electrolyte, has a chemical structure that only allows H^+ ions,
not the electrons, to pass through. The released electrons have to find their way
through the external electric circuit and reach the cathode side. This is the simplest
picture of a PEM's components and operation. Below, the various parts of PEMFCs
are discussed in more detail.

The catalyst electrodes consist of small grains of platinum, or other noble metals,
or their alloys that have a diameter of about 3 nm. These grains are supported by
carbon particles of around 50 nm in diameter. The grained structure provides the
highest catalytic surface, thus it increases the active area for chemical reactions to
take place. The carbon support ensures an electronic pathway between the catalyst
surfaces and the outer current collector. The contact of the inlet gases with the catalyst is provided by a special porous structure called the Gas Diffusion Layer (GDL),
shown in Figure 1.3. This porous structure allows the easy access of the inlet gases as

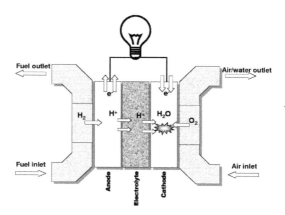

Figure 1.2: Schematic of a Proton Exchange Membrane (PEM) fuel cell.

well as the easy removal of the produced water from the cell. The GDL is impregnated by Poly-Tetra-Fluoro-Ethylene (PTFE) to become hydrophobic and prevent flooding of the cathode with water at high current densities.

The membrane of most PEMFCs are made of Nafion, a perfluorosulfonic acid membrane developed by E.I. DuPont de Nemours & Co. Nafion has excellent mechanical strength, water insolubility, and chemical and thermal stability. While the main body of Nafion is hydrophobic, its sulfonated side ends, shown in Figure 1.4, are hydrophilic such that long water chains can form inside the porous structure. This feature enables Nafion to have a high proton conductivity, allowing cations to exchange between the electrodes.

Several models have been proposed to describe the ionic behavior of Nafion membranes, for instance, equilibrium ionic selectivity and ionic transport properties [36, 44, 66, 115]. The properties of the membrane are considerably influenced by electrostatic forces between the aggregated clusters that are formed by ionic groups [24].

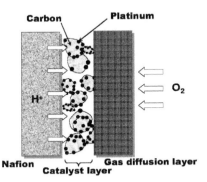

Figure 1.3: Illustrative structure of the cathode of a PEM fuel cell.

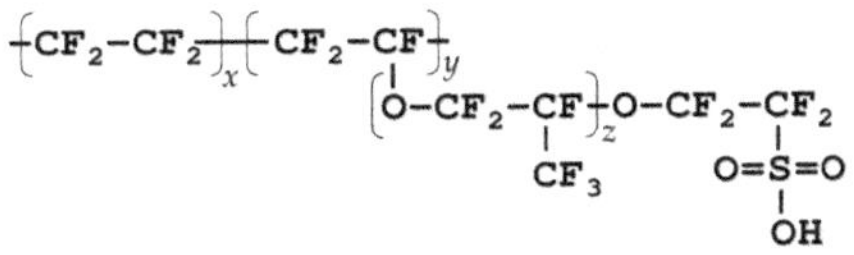

Figure 1.4: Structure of Nafion, the sulfonic ends can sustain water chains, while the remaining parts are hydrophobic.

Although these clusters are clearly observed in experiments, their morphology has not yet been fully understood.

In the Yeager and Steck model for Nafion, it is described by three regions: a fluorocarbon region (A), an interfacial zone (B) and an ionic cluster region (C). The fluorocarbon backbone (A) is highly hydrophobic, while region (C) is highly hydrophilic due to its dangling sulfonate groups. Thus, most absorbed water and counter-ions exist in region (C). The ionic clusters have been proposed to be spherical, forming a network interconnected by smaller channels [51]. The interfacial region (B) contains the side chains and sulfonate groups that are not forming clusters. Thus region (B) includes the counter-ions and only part of the absorbed water.

The function of Nafion as the membrane of a PEMFC is determined by its proton conductivity and water adsorption/removal properties. Its proton conductivity depends on the hydration number [2], which is a measure of the water/humidity content of Nafion. If the humidity falls below a certain value, Nafion behaves as a poor ion conductor. On the other hand, the ionic conductivity increases sharply with an increase in the water content. To avoid dryness, reactant gases are often humidified before they are fed into the GDL. This humidification technique can however, lead to the flooding phenomenon. Flooding refers to the over-presence of water in the electrode pores and the flow channels, preventing the reactant gases from reaching the active areas, especially at the cathode side. A good performance is often obtained when the humidity at the cathode is within a delicate range of 60-80%. Keeping the humidity within such an optimized range to prevent both dryness and flooding is denoted by water management issues in PEMFCs.

The cathodic electrochemical reactions that are vital to the operation of PEM fuel cells do not occur in all regions, but only at the so-called three-phase boundaries. At these reaction zones, the three components; namely the catalyst (solid), the electrolyte (liquid), and oxygen (gas) are simultaneously present. This ensures both the ionic and electronic paths required for the electrochemical reactions. The chemistry of the three-phase boundaries is therefore coupled to the transport of gases and the dynamic behavior of the electrolyte solution within a PEM fuel cell.

1.3 Multi-Scale Modeling of PEM Fuel Cells

The performance of a fuel cell, like any other battery, is best described by its current-versus-voltage (VI) curve. From a VI curve, one can determine the output characteristics of the cell such as its efficiency, load behavior, and power. The VI curve is obtained either by experiment, or by a model for the fuel cell.

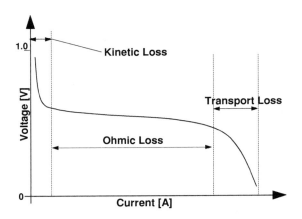

Figure 1.5: Schematic of a PEM fuel cell voltage versus current.

The typical VI curve of PEM fuel cells is shown in Figure 1.5. At very low current densities, the cell voltage drops substantially, attributed to the slow kinetics of the cell chemistry. At high voltages, the current goes to zero because of the flooding issue. Between those working regimes, the cell behaves similar to a normal battery. The initial drop in voltage is commonly referred to as the activation overpotential, which negatively affects the performance throughout the whole operational range. If the overpotential can be reduced by improving the charge transfer kinetics, the output power of the cell will increase substantially.

The chemistry at the anode, *i.e.*, the adsorption and oxidation of hydrogen, is much faster than that at the cathode. The chemistry at the cathode includes the adsorption of oxygen on the electrocatalyst surface, the diffusion of adsorbed oxygen-containing species, the hydrogen diffusion along the catalyst surface and the three-phase boundaries, and the charge transfer reactions.

In the current fuel cell models, the chemical processes are coupled to the transport and flow processes by the mean field (MF) method [59]. In the MF approach, one solves ordinary differential equations for the mean coverage of main surface species. Obviously, this method fails to simulate the complex chemistry at atomistic levels that may arise from a particular surface structure or its local topology. Neglecting the details of real chemistry has been repeatedly shown to generate incorrect predictions of the electrochemistry in fuel cells [116].

Multi-scale models can be proposed for PEM fuel cells, in which not only transport and flow, but also chemistry is simulated by the best available computational tools [87]. The core of our model for chemistry is based on the dynamic Monte Carlo (DMC) simulations, where instead of solving differential equations to simulate the evolution of the catalyst surface, one locally executes chemical reactions. This way, one can statistically follow the change in species coverage at a very detailed level. As a result, it becomes possible to capture the effect of surface structures using the DMC approach. Below, the DMC method and its dependence on the chemical rate data is discussed in more detail.

1.4 Dynamic Monte-Carlo Method and Kinetics

As mentioned in the previous section, the Dynamic Monte-Carlo (DMC) method is an efficient way to model random processes on catalyst surfaces [47, 53]. In DMC simulations, surface phenomena are divided into different categories, e.g., adsorption, diffusion, desorption and other chemical reactions. Adsorption results from the accumulation of gas molecules on the surface of a catalyst at both electrodes of a fuel cell. Hydrogen molecules can reside on the catalyst surface at the anode side and dissociate into separate H-atoms. In this way, the H-H bonds can be broken and new bonds can be created between single H atoms and catalyst atoms. On the cathode side, a similar but more complex kind of adsorption can occur for oxygen molecules.

Adsorbed atoms can then travel from one position on the surface to another position, representing the surface diffusion of atoms. The reverse process of dissociation, *i.e.*, re-combination, can also happen if two adsorbed atoms meet each other and form a bond. The product of this reaction can diffuse on the surface. On the other hand, if the new molecule goes back to the gas phase, the process is that of desorption. It can be seen that all different surface phenomena can be defined as the occupation or evacuation of certain positions on the surface, favored positions which are called sites.

The definition of surface sites allows one to represent a crystal surface by a discrete number of points where adsorbed atoms can sit. By numbering all available sites on a given surface, it is possible to trace the occupation and evacuation of surface sites. Generally, every process can be split into single discrete events. Each event describes the occupation or evacuation of one or more sites or a change in the adsorbates. Therefore, one can follow surface reactions as consequent random events occurring one at a time. Changing the configuration of surface sites depends on the transition probability values assigned to each process or each class of events.

In DMC simulations, a single reaction event is chosen at a random time to change the configuration of the system. The event selection step depends on the probability with which the events can occur. This probability, called the transition probability, is the microscopic equivalent to the reaction rates. For a reaction that macroscopically happens at a higher reaction rate, the microscopic single reaction event occurs at a higher transition probability. Reaction rates also influence the time advancement of the Dynamic Monte-Carlo method. The faster the reactions occur, the smaller the time step will be. As a result, the Dynamic Monte-Carlo method will record more changes of the system within a fixed amount of simulation time.

By construct, DMC simulations require the chemical kinetic rate data that corresponds to the chemistry of fuel cells. The kinetic rate data determines the probability of how fast each reaction can happen on the surface. The more accurate the rate constants for each reaction, the more successful DMC will be in predicting the chemistry. Nevertheless, even if the rates lack absolute precision, DMC simulations can still generate reliable results, provided that the ratio of rates is accurate enough. Therefore in this work, special attention is paid to the consistency of the chemical kinetic rate data.

1.5 General Form of the Rate Constants

Rate constants are generally given in the Boltzmann Arrhenius form:

$$k = AT^b \exp\left(-\frac{\Delta E}{k_B T}\right) \tag{1.3}$$

or

$$k = A \exp\left(-\frac{\Delta E}{k_B T}\right) \tag{1.4}$$

where:
- k reaction coefficient,
- A pre-exponential factor constant that does not depend on temperature,
- T absolute temperature in Kelvin,
- b temperature exponent,
- ΔE energy barrier of the reaction,
- k_B Boltzmann constant ($1.3806503 \times 10^{-23}$ m^2 kg s^{-2} K^{-1})

To determine a rate, one has to obtain at least two parameters, namely the pre-exponential factor, A and the energy barrier, ΔE. These two values are tabulated in the literature, where reaction rates are to be presented. Rates of all kinds of reactions, except the adsorption process, are given in this form.

For adsorption reactions, usually only a single quantity, the sticking coefficient, is reported. As a gas molecule bumps onto a solid surface, it may be captured by surface atoms. However, due to the kinetic energy of the molecule, not all collisions lead to adsorption. The ratio of the number of adsorbed molecules to the total number of collisions is called the sticking coefficient. Its value depends on temperature as well as the surface coverage of adatoms. As the temperature rises, the probability of adsorption reduces since gas molecules have a higher kinetic energy and can bounce back from the surface more easily. This way, the sticking coefficient usually decreases as the temperature increases. The sticking coefficient also depends on the coverage of adatoms. Usually, the term *surface coverage* refers to the ratio of the adsorbed atoms to the total number of the available surface sites. As the coverage of adsorbed atoms increases, less collisions can lead to adsorption. In summary, the rate of adsorption is determined by finding the sticking coefficient, which can be further related to the corresponding rate constant of adsorption.

For other kinds of reactions, generally two parameters have to be computed: the pre-exponential factor and the energy barrier. Any reaction is represented by a change of the system from an initial state to a final state. The energy barrier associated with this process is shown in Figure 1.6.

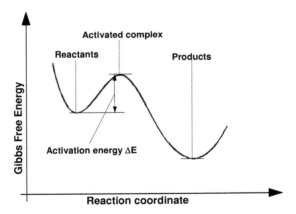

Figure 1.6: Change in the Gibbs free energy during a reaction. The activated complex is the configuration of maximum Gibbs energy along the reaction path that connects the reactants to the products.

1.6   Rate Constants from the Literature

As mentioned above, rate parameters for different surface reactions are specified by reporting two quantities: the pre-exponential factor, A and the energy barrier, ΔE. For adsorption processes, a single parameter called the sticking coefficient, determines the corresponding adsorption rate. Historically, these parameters have been obtained through experiments. Some experimental data are available for the surface reactions of the ORR, but the conditions (e.g. temperature) under which they were obtained vary. The best reference for these reactions has been compiled by Försth [41]. Table 1.1 presents all the available parameters for the surface reactions of hydrogen and oxygen on platinum, as reported by Försth.

Note that the pre-exponential factors have been treated in a rather simple way. For example, the same value 3.7×10^{21} is reported for the desorption of H_2, O_2, and all the Langmuir-Hishelwood surface reactions. The activation energies have been modified to better match the experimental measurements and the low-coverage calculations [41]. Moreover, these data have been obtained by different research groups in different experiments. Therefore, they lack compatibility, a characteristic required by the DMC method to generate meaningful results. An alternative way to determine rate data is through quantum mechanical calculations, where reactions can be simulated in a consistent way. The topic of this book is the application of quantum calculations to obtain kinetic parameters of electrochemical reactions in PEM fuel cells.

1.7   Rate Constants and Quantum Simulations

Quantum or ab-initio methods generally refer to computations that simulate the atoms and ions present in a system without using experimentally fitted parameters. Application of quantum mechanics to study atomic and electronic behavior of materials dates back to two decades ago [84]. From very early comparisons between quantum

Table 1.1: Rate constants for surface reactions, Reference Försth [41]

Reaction	A	ΔE [J mol^{-1}]
Adsorption and desorption reactions		
$H_2 \Rightarrow 2H_{ads}$	0.046 (sticking)	-
$2H_{ads} \Rightarrow H_2$	3.7×10^{21}	67000
$O_2 \Rightarrow 2O_{ads}$	0.023 (sticking)	-
$2O_{ads} \Rightarrow O_2$	3.7×10^{21}	213384
Langmuir-Hishelwood surface reactions		
$H_{ads}+O_{ads} \Rightarrow OH_{ads}$	3.7×10^{21}	54340
$H_{ads}+OH_{ads} \Rightarrow H_2O_{ads}$	3.7×10^{21}	64636
$OH_{ads}+OH_{ads} \Rightarrow H_2O_{ads}+O_{ads}$	3.7×10^{21}	74283
Adsorption and desorption		
of products and intermediates		
$H_2O \Rightarrow H_2O_{ads}$	0.7(sticking)	-
$H_2O_{ads} \Rightarrow H_2O$	10^{13}	65000
$OH \Rightarrow OH_{ads}$	1.0 (sticking)	-
$OH_{ads} \Rightarrow OH$	10^{14}	245039
$O \Rightarrow O_{ads}$	1.0 (sticking)	-
$O_{ads} \Rightarrow O$	10^{13}	356000
$H \Rightarrow H_{ads}$	1.0 (sticking)	-
$H_{ads} \Rightarrow H$	10^{13}	249000

simulations with experimental results, it became obvious that modern quantum theory can often predict physical phenomena with good accuracy. For instance, one can mention the prediction of the lattice constants of solid crystals, covalent bond energies, and the distinction between insulators and conductors based on quantum simulations. There is no counter-fact to the application of quantum theory and its failure has yet to be shown.

In the early quantum calculations, it was only possible to model systems comprising of a few atoms, calculations which were limited to very small systems. Appreciable improvements in both computer power as well as efficient mathematical methods, have made it possible now to simulate systems composed of about one hundred atoms within a reasonable time. Computing the total energy, its changes and derivatives can provide almost all physical properties of interest for a given system. As an example, the equilibrium lattice constant of a crystal is the value that minimizes the total energy of the crystal. Therefore, to find the lattice constant, one sets up several quantum simulations for the same crystal, each having a different lattice constant as a parameter. Then, the best estimate of the lattice constant is determined by the point on the energy curve that gives the lowest energy.

Energy computations can be also used to calculate kinetic data. If E_1 and E_2 denote the total energy of the system in the initial and the transition states, the energy barrier ΔE is defined by:

$$\Delta E \;=\; E_2 - E_1 \,. \tag{1.5}$$

If the total energies can be obtained from quantum simulations, it is possible to find the energy barrier of a reaction. Hence, simulation steps can be easily summarized as follows:

1. Define atomic structure of the initial and the transition state of the system.

2. Perform quantum simulations to get the total energy of the system in both states.

3. Subtract these energies to get the energy barrier.

The initial state denotes a local minimum of the total energy, while the transition state refers to a local maximum of the energy along the reaction coordinate (Figure 1.5). The other parameter required to compute a rate is the pre-exponential factor A in equation (1.4). Roughly speaking, A represents the frequency with which the system oscillates in its initial state. This frequency is usually determined from a statistical mechanics analysis of the reactants in their initial local minimum on the potential energy surface (PES). All possible oscillations of a system can affect this frequency, thereby complicating its determination.

If the transition from the initial to the final state can be approximated by a single degree of freedom, the motion of only one atom characterizes the state of the system in that reaction. Then, one can analyze such a one dimensional motion within the Harmonic model to estimate the frequency factor. When all the degrees of freedom are taken into account, one computes the full second derivative of the total energy, the Hessian matrix, instead of only one scalar value. The Hessian matrix is subsequently diagonalized, and its eigenvalues are used in a way similar to the one dimensional case. Thus, by computing the total energy of the system, one not only can estimate the energy barrier of a reaction but also determine the pre-exponential factor of the rate equation.

In this work, quantum mechanical computations are used to simulate various chemical processes in PEM fuel cells. Chapter 2 is a brief review of the fundamental theories and computational approaches in quantum chemistry. These theories adapt the Schrödinger equation, from its original physical form, to modeling atomistic systems as those found in chemistry. Chapter 3 first reviews the oxygen reduction reaction (ORR) mechanism on platinum surfaces, and then focuses on the treatment of electron transfer reactions from both macroscopic and microscopic points of view.

In Chapter 4, the *Local Reaction Center* (LRC) theory based on the Marcus electron transfer (ET) model is discussed. The LRC theory establishes a computational framework to investigate electrochemical processes at the electrodes and the corresponding transition states. On the catalytic surfaces, the activation energy of an ET process is an explicit function of the electrode potential. This additional complexity increases the computational cost to find the transition states of ET reactions. Therefore, an efficient search algorithm that facilitates determining the transition states for large systems within the LRC theory is developed. Chapter 5 investigates the computational details of the LRC theory in its current form, and monitors the reasons of those limitations.

Chapter 6 shows the application of the methods, discussed in Chapter 4, to ORR mechanisms on platinum surfaces. In Chapter 7, a mathematical framework to estimate the pre-exponential factors of ET reactions is derived within the Harmonic approximation, and using quantum calculations. In Appendix B, it is proved that pre-exponential factors of ET reactions for oxidation and reduction steps are equal, at least under standard and equilibrium conditions. Appendix A explores two ideas in extending the LRC theory to be applied to systems with many catalyst atoms. Useful discussions and results in the appendix provide valuable insights for future applications of quantum calculations in the field of electrochemistry.

Chapter 2

Computational Quantum Chemistry

Computational quantum chemistry refers to a branch of chemistry that, similar to Molecular Dynamics (MD), attempts to predict the structure, properties and activity of materials using computer simulations. While in MD, atoms are modeled as rigid bodies; in quantum chemistry, atoms are decomposed into nuclei and electrons, which are modeled by the Schrödinger equation as waves. Even if all the nuclei are treated as classical objects, this equation can be analytically solved only for very special molecules, those which are hydrogen-like and have no more than one electron. Since almost all the systems of interest have many electrons, the Schrödinger equation has to be solved by numerical methods. Hence in almost all cases, one has to resort to approximate solutions.

To handle practical systems, three components are required. First, model theories are needed to approximate the solution up to desired accuracies. Second, basis sets and/or pseudopotentials are used to transfer the single-body electronic information from the atomic environment to the many-body electronic environment of molecules. Finally, mathematical techniques are applied to solve the already derived equations in the most effective ways. Only the first two components are covered in this chapter.

The model theories range from purely ab-initio ones, those which need no experimental parameters, to empirical models, those which are fitted models to a certain set of experimental data. Between these two classes, there exist numerous schemes that combine the two types to achieve better performance or higher accuracy or both. As for the second aspect, atomic-based basis functions and plane waves are two main approaches that are accordingly used for non-periodic, and periodic systems. At present, no single approach is suitable for all applications, but only to certain classes of problems. Developing and testing new basis sets is yet another field that serves computational chemistry.

The goal of this chapter is to present the fundamental theories and methods on which modern quantum calculations rely. Mathematical formulations are presented in a concise but consistent way to solidify the necessary background, and illustrate the concepts used in the next chapters. This chapter by no means claims to be a complete review of all the topics in the field. Its purpose is to cover the main benchmarks, and to pave the path for the application of quantum calculations used widely in this book. All the materials are based on the standard textbooks in computational chemistry, and in particular on references [30, 31, 43, 54, 79, 98]. Whenever possible or necessary, reference has been made to relevant publications in the literature.

2.1 Schrödinger Equation

In quantum mechanics, the time-dependent position of a particle is determined by a complex function $\Psi(\mathbf{x}, t)$, whose magnitude-squared equals the probability density of finding the particle at position $\mathbf{x}$ and at time t. The relationship between the wave function $\Psi(\mathbf{x}, t)$, and the probability distribution is mathematically an integral over subspace Ω [30]

$$P = \int_\Omega \Psi(\mathbf{x}, t)^* \Psi(\mathbf{x}, t) d\mathbf{x} \qquad (2.1)$$

where $\Psi(\mathbf{x}, t)^*$ is the conjugate of $\Psi(\mathbf{x}, t)$. By definition, if Ω extends over all the space, the probability in equation (2.1) should be equal to unity, that is

$$1 = \int_{-\infty}^{+\infty} \Psi(\mathbf{x}, t)^* \Psi(\mathbf{x}, t) d\mathbf{x} \tag{2.2}$$

Equation (2.2) simply declares that the particle does exist within the whole space; no matter how its position probability is distributed. Despite its simple form, constraint (2.2) on the wave function plays an important role when any quantum problem is solved. It often enters into the formulation of a problem through the Lagrange multipliers method.

$\Psi(\mathbf{x}, t)$ for any particle or system can be obtained by solving the *time-dependent* Schrödinger equation

$$\hat{\mathbf{H}} \Psi(\mathbf{x}, t) = i\hbar \frac{\partial}{\partial t} \Psi(\mathbf{x}, t) \tag{2.3}$$

where $\hat{\mathbf{H}}$ is the Hamiltonian operator, and $\hbar \equiv \frac{h}{2\pi} \cong 1.05457 \, 10^{-34}$ J.s is the *reduced Planck's constant*. Details about the form of operator $\hat{\mathbf{H}}$ for chemical systems will be discussed in the next section. Here, we focus on the general solution of equation (2.3). According to the postulates of Quantum Mechanics, the solution $\Psi(\mathbf{x}, t)$ of equation (2.3) contains all the information for the considered system. Once $\Psi(\mathbf{x}, t)$ is known, any physical quantity A can be estimated through the expected value integral

$$< A(t) > \equiv \frac{\int_{-\infty}^{+\infty} \Psi(\mathbf{x}, t)^* \hat{\mathbf{A}}(\mathbf{x}, t) \Psi(\mathbf{x}, t) d\mathbf{x}}{\int_{-\infty}^{+\infty} \Psi(\mathbf{x}, t)^* \Psi(\mathbf{x}, t) d\mathbf{x}} \tag{2.4}$$

where $\hat{\mathbf{A}}(\mathbf{x}, t)$ is a Hermitian operator associated with quantity A. For example, the total energy of the system is equal to the expected value of the Hamiltonian $\hat{\mathbf{H}}$. The denominator in equation (2.4) is introduced in the definition to assure the normalization constraint (2.2) for non-normalized wave functions.

The time-Dependent Schrödinger equation (2.3) is a partial differential equation in terms of two types of independent variables: spatial coordinates $\mathbf{x}$, and time t. If the Hamiltonian is independent of time, it is possible to separate the dependency of the wave function on these two variables. Fortunately, this condition is satisfied for most physical systems. Assuming a separable solution in terms of time and space for the wave function, one can write

$$\Psi(\mathbf{x}, t) = \Psi_{\mathbf{x}}(\mathbf{x}) \cdot \Psi_t(t) \tag{2.5}$$

Substitution of this form in equation (2.3) leads to

$$\frac{\hat{H}\Psi_{\mathbf{x}}(\mathbf{x})}{\Psi_{\mathbf{x}}(\mathbf{x})} = i\hbar \frac{\frac{\partial \Psi_t(t)}{\partial t}}{\Psi_t(t)} \tag{2.6}$$

The right hand side of this equation is a function of time only, while the left hand side depends only on space coordinates. This implies that each side is equal to a constant, namely E with the dimensions of energy

$$i\hbar \frac{\frac{\partial \Psi_t(t)}{\partial t}}{\Psi_t(t)} = E \tag{2.7}$$

$$\frac{\hat{H}\Psi_{\mathbf{x}}(\mathbf{x})}{\Psi_{\mathbf{x}}(\mathbf{x})} = E \tag{2.8}$$

Equation (2.7) can be readily solved, which gives

$$\Psi_t(t) = \Psi_{t0} \exp\left(-\frac{iE}{\hbar}t\right) \tag{2.9}$$

where Ψ_{t0} is the constant of integration, being equal to the initial value of $\Psi_t(t)$. The exponential part in equation (2.9) introduces only a phase-shift in $\Psi(\mathbf{x}, t)$, but has no effect on its magnitude. This can be shown by substituting solution (2.9) in equation (2.5), and evaluating the magnitude of $\Psi(\mathbf{x}, t)$

$$
\begin{aligned}
|\Psi(\mathbf{x}, t)|^2 &= |\Psi_\mathbf{x}(\mathbf{x})|^2 \cdot |\Psi_t(t)|^2 \\
&= |\Psi_\mathbf{x}(\mathbf{x})|^2 \cdot \left[\Psi_{t0} \exp\left(-\frac{iE}{\hbar} t \right) \right]^* \left[\Psi_{t0} \exp\left(-\frac{iE}{\hbar} t \right) \right] \\
&= |\Psi_\mathbf{x}(\mathbf{x})|^2 \cdot |\Psi_{t0}|^2 \cdot \left[\exp\left(+\frac{iE}{\hbar} t \right) \exp\left(-\frac{iE}{\hbar} t \right) \right] \\
&= |\Psi_\mathbf{x}(\mathbf{x})|^2 \cdot |\Psi_{t0}|^2
\end{aligned}
\tag{2.10}
$$

Furthermore, Ψ_{t0} can be absorbed into $\Psi_\mathbf{x}(\mathbf{x})$ because the solution of equation (2.8) will remain unchanged if it is multiplied by any non-zero constant. In this situation, one may drop the subscript $\mathbf{x}$ from $\Psi_\mathbf{x}(\mathbf{x})$ and rewrite equation (2.8) to arrive at the *time-independent* Schrödinger equation

$$
\hat{\mathbf{H}} \Psi(\mathbf{x}) = E \Psi(\mathbf{x})
\tag{2.11}
$$

This is still a partial differential equation in terms of the $3N$ space coordinates of N particles present in the system. After decades since the formulation of the Schrödinger equation, analytical solutions for most systems cannot be found, and even approximate numerical solutions involve a large amount of mathematical and computational complexity. For chemical systems, the complexity can be reduced by assuming that the motion of nuclei is separable from that of the electrons. Suggested by Born and Oppenheimer (BO) in 1927 [23], this idea is the basic assumption of almost all modern quantum calculations of atoms and molecules. The BO approach, despite its approximate nature, has been quite successful in predicting the properties of chemical systems.

2.2   Born-Oppenheimer Approximation

Born and Oppenheimer [23] showed that the total wave function of a chemical system could be well approximated by the product of a nuclear wave function Ψ_{nuc} and an electronic wave function Ψ_{elec}. Ψ_{nuc} depends only on the nuclei positions $\mathbf{R} = \{\mathbf{R}_I\}$, while Ψ_{elec} depends on both nuclear, and electronic coordinates $\mathbf{r} = \{\mathbf{r}_i\}$ [30].

$$\Psi_{sys}(\mathbf{R}, \mathbf{r}) = \Psi_{nuc}(\mathbf{R}) \cdot \Psi_{elec}(\mathbf{R}, \mathbf{r}) \qquad (2.12)$$

The benefit of this separated form is Ψ_{nuc} and Ψ_{elec} can now be obtained from two simpler, though coupled, Schrödinger equations

$$\left[\hat{\mathbf{H}}_{nuc} + E_{elec}\right] \Psi_{nuc}(\mathbf{R}) = E_{nuc}\Psi_{nuc}(\mathbf{R}) \qquad (2.13)$$

$$\hat{\mathbf{H}}_{elec}\Psi_{elec}(\mathbf{R}, \mathbf{r}) = E_{elec}\Psi_{elec}(\mathbf{R}, \mathbf{r}) \qquad (2.14)$$

In equation (2.13), the Hamiltonian is a function of electronic energy E_{elec}. On the other hand, in equation (2.14), E_{elec} is a part of the solution that depends on $\mathbf{R}$ as a part of input parameters. $\hat{\mathbf{H}}_{nuc}$ contains all non-electronic energies: nuclei repulsion, nuclei vibrations, and translational as well as rotational energies of the whole molecule. In computational chemistry, these components of the total energy are often estimated classically, in the sense that they are evaluated by treating the nuclei as rigid objects. An extensive statistical mechanics treatise of the classical energy components can be found in Reference [67]. The quantum mechanical nature of the problem is thereby localized in the electronic energy component, that is, the eigenvalue problem (2.14). The electronic energy E_{elec} is the expected value of the Hamiltonian operator $\hat{\mathbf{H}}_{elec}$. Depending parametrically on $\mathbf{R}$ for a system of N_n nuclei and N_e electrons, $\hat{\mathbf{H}}_{elec}$ is given by [98]

$$\hat{\mathbf{H}}_{elec} = \sum_{i=1}^{N_e} \left(-\frac{1}{2}\nabla_i^2\right) - \sum_{i=1}^{N_e}\sum_{I=1}^{N_n} \frac{Z_I}{|\mathbf{R}_I - \mathbf{r}_i|} + \sum_{i=1}^{N_e}\sum_{j>i}^{N_e} \frac{1}{|\mathbf{r}_i - \mathbf{r}_j|} \qquad (2.15)$$

Equation (2.15) is in atomic units, where Planck's constant h, vacuum permittivity $4\pi\epsilon_o$, electronic mass m_e, and electronic charge e are all assigned a value of unity. In the first summation term, ∇_i^2 is the Laplacian operator corresponding to the kinetic energy. Subscript i indicates that it acts only on the coordinates of atom i. Using a

Cartesian coordinate system $\mathbf{r}_i = (x_i, y_i, z_i)$, ∇_i^2 equals

$$\nabla_i^2 = \frac{\partial^2}{\partial x_i^2} + \frac{\partial^2}{\partial y_i^2} + \frac{\partial^2}{\partial z_i^2} \tag{2.16}$$

The second and third summation terms in equation (2.15) represent the Coulombic attraction between the nuclei and the electrons, and the repulsion between the electrons, respectively. Z_I stands for the atomic number of nucleus I. The BO approximation allows one to treat the non-electronic energies classically, and calculate the quantum portion of energy E_{elec} from equation (2.15) as a function of the nuclei configuration

$$E_{elec} = E_{elec}(\mathbf{R}). \tag{2.17}$$

Therefore, instead of treating $N_n + N_e$ particles quantum mechanically by solving the Schrödinger wave equation (2.11), it is actually solved only for the electrons, and not for the nuclei.

At this point, it is beneficial to discuss the rationale for the BO approximation based on physical considerations. If one assumes a moderately uniform distribution of kinetic energy between the particles within a molecule, then according to the *equipartition* law [30], electrons and nuclei must possess the same amount of kinetic energy. In this situation, the ratio of their velocities is proportional to the inverse square root of their mass ratio. On average, a nucleus mass is 10,000 times the mass of an electron, which means that a typical electron moves roughly 100 times faster than a nucleus. From the point of view of the electrons, the nuclei are fixed objects imposing a steady field of attraction force. On the other hand, the nuclei see the electrons not as distinguishable objects, but rather similar to 3-dimensional clouds with a negative charge. Electronic clouds adjust themselves almost instantly to any configuration of the nuclei. Thus it seems quite justifiable to attribute the quantum effects to the electrons only.

Although the BO formulation (2.15) is used in almost all quantum chemical computations, one must be aware of the involved limitations mentioned before. Moreover, the Schrödinger equation (2.3) lacks two fundamental features of modern quantum physics. First, the electrons are considered as having fixed masses, while according to the relativity theorem, their mass depends on their velocity. In particular, the relativity effect becomes significant for the inner most electronic shells, *i.e.*, the *core* electrons in heavy atoms. Fortunately, this effect can be captured to a good extent by using appropriate basis functions and/or *pseudo-potentials*. At this point, these are just mathematical functions that can replace the core of heavy atoms in quantum calculations. For more sophisticated situations, one should appeal to the *relativistic quantum mechanical wave equation* formulated by Dirac in 1928.

The second shortcoming of equations (2.3) and (2.15) is their neglect of spin for electrons. The direct solution to those equations, if constructed without the required provisions, may contradict *Pauli's exclusion* principle. According to this principle, at any time no two electrons with the same spin in any atomic or molecular system can occupy the same location in space. Actually, a satisfactory solution can be constructed by requiring that the wave function be antisymmetric with respect to the exchange of any two electrons. In the next sections, it is shown that one way to enforce this condition is to use a special determinantal form, the Slater determinant, to set up an antisymmetric trial wave function. Slater determinants are the basic building blocks of all *Molecular Orbital* (or *Wave Function*) methods. Because any Slater determinant automatically satisfies the Pauli exclusion principle, any wave function if written in the form of a linear combination of such determinants, will also be antisymmetric.

2.3   Molecular Orbital (Wave Function) Models

2.3.1   Slater Determinants

For most molecules of interest, no analytical solution is known for equation (2.14), and it has to be solved by numerical methods. Similar to other branches of science, a numerical approach becomes more effective if it is developed based on the physical insights about the solution. One of the most successful numerical schemes in computational quantum chemistry is based on using *basis functions*, which describe the probability distributions of finding electrons in space when in isolated atoms. These probability functions are called atomic orbitals (AOs).

Using atomic orbitals to build a molecular wave function is the main concept of Molecular Orbital models. These models are also called Wave Function methods to emphasize the application of wave function as the principle quantity on which computational formulations are based. An orbital is a space function for an electron that, when occupied, determines the probability of finding the occupying electron in any given spatial region. It is common to indicate orbital ϕ_i has been occupied by an electron having coordinates $\mathbf{r}_j$ by writing $\phi_i(\mathbf{r}_j)$. An arbitrary distribution of electrons in a molecule thus can be represented by assigning individual electrons to a product of non-interacting orbitals

$$\phi_1(\mathbf{r}_{j_1}) \cdot \phi_2(\mathbf{r}_{j_2}) \cdot ... \cdot \phi_{N_e}(\mathbf{r}_{j_{N_e}}) \tag{2.18}$$

The product form in (2.18) lacks any information about the spin of the electrons. To include spin, electronic coordinate $\mathbf{r}$ is augmented by adding spin s as a new variable

$$\mathbf{x} \equiv (\mathbf{r}, s) \tag{2.19}$$

Next, two spin functions $\alpha(s)$ and $\beta(s)$ are defined such that they represent the direction of the electronic spin. The exact forms of $\alpha(s)$ and $\beta(s)$ are neither known, nor required for their application in mathematical derivations. Instead, they are characterized by their properties whens acted upon by quantum operators. In particular, spin functions α and β are orthonormal: any integral over spin equals either unity if taken over the same spin function, or zero if taken over different spin functions, *i.e.*

$$\int \alpha(s_1)\,\alpha(s_1)\,d\sigma_1 = \int \beta(s_1)\,\beta(s_1)\,d\sigma_1 = 1 \qquad (2.20)$$

$$\int \alpha(s_1)\,\beta(s_1)\,d\sigma_1 = \int \beta(s_1)\,\alpha(s_1)\,d\sigma_1 = 0 \qquad (2.21)$$

Spin-orbital φ is now defined as the product of *space-orbital* ϕ and either spin function α or β

$$\varphi(\mathbf{x}) \equiv \phi(\mathbf{r}) \cdot \begin{cases} \alpha(s) \\ \text{or} \\ \beta(s) \end{cases} \qquad (2.22)$$

Finally, the product configuration (2.18) is rewritten in terms of spin-orbitals $\{\varphi_i\}$, and augmented coordinates $\{\mathbf{x}_j\}$ as

$$\varphi_1(\mathbf{x}_{j_1}) \cdot \varphi_2(\mathbf{x}_{j_2}) \cdot \ldots \cdot \varphi_{N_e}(\mathbf{x}_{j_{N_e}}) \qquad (2.23)$$

At this point, two valid questions are how to distribute N_e electrons among N_e spin orbitals, and how to construct a trial wave function. For now, we consider only fully occupied or totally empty orbitals. Because the electrons are indistinguishable from each other, there must be no preference as which electron occupies which orbital. Therefore, for any given set $\{\varphi_i, i = 1...N_e\}$, there exist $N_e!$ number of possible configurations. For example, one can write six product configurations to assign three electrons to three orbitals

$$
\begin{aligned}
&\varphi_1(\mathbf{x}_1)\,\varphi_2(\mathbf{x}_2)\,\varphi_3(\mathbf{x}_3) \\
&\varphi_1(\mathbf{x}_1)\,\varphi_2(\mathbf{x}_3)\,\varphi_3(\mathbf{x}_2) \\
&\varphi_1(\mathbf{x}_2)\,\varphi_2(\mathbf{x}_1)\,\varphi_3(\mathbf{x}_3) \\
&\varphi_1(\mathbf{x}_2)\,\varphi_2(\mathbf{x}_3)\,\varphi_3(\mathbf{x}_1) \\
&\varphi_1(\mathbf{x}_3)\,\varphi_2(\mathbf{x}_1)\,\varphi_3(\mathbf{x}_2) \\
&\varphi_1(\mathbf{x}_3)\,\varphi_2(\mathbf{x}_2)\,\varphi_3(\mathbf{x}_1)
\end{aligned}
\tag{2.24}
$$

One way to create an electronic wave function out of such products is to form a determinant, known as the Slater determinant

$$
\Psi = \frac{1}{\sqrt{3!}}
\begin{vmatrix}
\varphi_1(\mathbf{x}_1) & \varphi_1(\mathbf{x}_2) & \varphi_1(\mathbf{x}_3) \\
\varphi_2(\mathbf{x}_1) & \varphi_2(\mathbf{x}_2) & \varphi_2(\mathbf{x}_3) \\
\varphi_3(\mathbf{x}_1) & \varphi_3(\mathbf{x}_2) & \varphi_3(\mathbf{x}_3)
\end{vmatrix}
\tag{2.25}
$$

where the rows denote orbitals, and the columns represent the electrons. Factor $\frac{1}{\sqrt{3!}}$ is to normalize Ψ. Form (2.25) can be easily generalized to an $N_e \cdot N_e$ determinant. The $N_e \cdot N_e$ Slater determinant is given here for future reference.

$$
\Psi = \frac{1}{\sqrt{N_e!}}
\begin{vmatrix}
\varphi_1(\mathbf{x}_1) & \varphi_1(\mathbf{x}_2) & .. & \varphi_1(\mathbf{x}_{N_e}) \\
\varphi_2(\mathbf{x}_1) & \varphi_2(\mathbf{x}_2) & .. & \varphi_2(\mathbf{x}_{N_e}) \\
. & . & .. & . \\
. & . & .. & . \\
\varphi_{N_e}(\mathbf{x}_1) & \varphi_{N_e}(\mathbf{x}_2) & .. & \varphi_{N_e}(\mathbf{x}_{N_e})
\end{vmatrix}
\tag{2.26}
$$

Slater determinants, by construct, satisfy the Pauli exclusion principle. If any two columns that correspond to two electrons are exchanged, the sign of the wave function Ψ changes.

For now, it has been assumed that only N_e spin orbitals are available. In practice, one may choose N_e orbitals from a given set with $N > N_e$ orbitals, and fill each orbital with an occupation number between zero and one. This way, different

Slater determinants corresponding to different choices may be formed. The total wave function Ψ can then be written as a linear combination of such Slater determinants

$$\Psi = \sum_{K=1} D_K \Psi_K \qquad (2.27)$$

where $\{D_K\}$ are the coefficients of the Slater determinants $\{\Psi_K\}$. $\{D_K\}$ are found by applying the *variation principle* described below. The variation (or variational) principle provides a direct way to optimize any candidate solution. It is used in almost all quantum theories to derive the desired computational schemes.

2.3.2 Variation Principle

Suppose that a trial wave function $\tilde{\Psi}$ has been constructed based on a parametric combination of functions belonging to a Hilbert space. The variation principle enables us to determine those parameters such that $\tilde{\Psi}$ be optimized, *i.e.*, it becomes the best possible answer for the solution of the time-independent Schrödinger equation (2.11). The elements of the Hilbert space can be, for example, one electron orbitals φ in (2.22), or Slater determinants as in (2.26). The elements or the basis functions can be combined in a non-linear way, or linearly as in equation (2.27).

The variation principle states that the best solution is the one for which the estimated energy $\tilde{E}$ of the system is minimized, thereby the variation of $\tilde{E}$ should vanish

$$\delta\tilde{E} = 0 \qquad (2.28)$$

Equation (2.28) can be used to obtain the unknown parameters appearing in the construction of $\tilde{\Psi}$. It simply states that the derivatives of energy with respect to those parameters vanish at the optimum wave function. The estimated energy obtained this way will always be higher than or equal to the ground state energy.

$$\tilde{E} \leq E_{\text{exact}} \qquad (2.29)$$

The equality case is expected to occur only when the set of basis functions contains an infinite number of elements. As using such a set is practically impossible, the ground state energy is the lower bound for all energies that are estimated by approximate variational methods. Hopefully, a rich set of functions is flexible enough to represent the exact wave function within a desired accuracy.

It should be noted that not every method to solve the Schrödinger equation is by default variational. In variational methods, the energy is always an upper-bound to the exact energy of the system. By systematic improvements of a variational method, if possible, the estimated value of energy monotonically decreases to its exact value. In non-variational approaches, such as perturbation-based methods, the calculated energy at each improvement step may be lower, higher or equal to the exact energy. In these cases, subsequent improvements allow the energy to converge to a value that could be exact.

Below, it is shown that for a linear combination of wave functions $\{\Psi_K\}$ as the basis set, the variation principle transforms the Schrödinger equation (2.11) into a generalized eigenvalue problem, which can be solved by standard routines in Linear Algebra.

Consider a trial electronic wave function $\tilde{\Psi}$ that is a linear combination of basis functions $\{\Psi_K\}$ as in (2.27). We substitute this linear form into equation (2.11)

$$\hat{\mathbf{H}} \sum_{K=1} D_K \Psi_K = \tilde{E} \sum_{K=1} D_K \Psi_K \, , \tag{2.30}$$

where the subscript *elec* has been dropped from $\hat{\mathbf{H}}$ and E for convenience. The expected value $\tilde{E}$ is determined from equation (2.4)

$$\tilde{E} = \frac{\sum_{K=1}\sum_{L=1} D_K^* D_L \int \Psi_K^* \hat{\mathbf{H}} \Psi_L d\mathbf{x}}{\sum_{K=1}\sum_{L=1} D_K^* D_L \int \Psi_K^* \Psi_L d\mathbf{x}} \, . \tag{2.31}$$

The integral elements in (2.31) can be readily evaluated because $\{\Psi_K\}$ are presumably known functions of electronic coordinates $\mathbf{x}$

$$\tilde{E} = \frac{\sum\limits_{K=1}\sum\limits_{L=1} D_K^* D_L H_{KL}}{\sum\limits_{K=1}\sum\limits_{L=1} D_K^* D_L S_{KL}} . \tag{2.32}$$

$\{H_{KL}\}$ are the elements of matrix $\mathbf{H}$, which is the representation of the Hamiltonian operator $\hat{\mathbf{H}}$ in the Hilbert space spanned by $\{\Psi_K\}$. $\{S_{KL}\}$ are the elements of the *overlap* matrix $\mathbf{S}$. $\mathbf{S}$ becomes the identity matrix if set $\{\Psi_K\}$ is an orthonormal basis set.

We assume here that $\{\Psi_K\}$ are fixed and cannot vary. Then to find the stationary value of $\tilde{E}$ according to equation (2.28), its derivatives should vanish with respect to each D_K

$$\sum_{L=1} D_L H_{KL} = \tilde{E} \sum_{L=1} D_L S_{KL} \quad \forall K \tag{2.33}$$

or in matrix form

$$\mathbf{HD} = \tilde{E}\mathbf{SD} \tag{2.34}$$

which is a generalized eigenvalue problem. Equation (2.34) can be solved for parameters $\{D_K\}$ and $\tilde{E}$, which are the only undetermined values. The main roadmap of Molecular Orbital methods is to form a trial wave function as in (2.27), and then optimize it using the variation principle. The more configurations $\{\Psi_K\}$ are used in the expansion of $\tilde{\Psi}$, the closer will the optimized solution be to the exact wave function. This constitutes the fundamental idea of one of the most accurate computational schemes, called the *configuration interaction* method. The simplest case of a single Slater determinant, the *Hartree-Fock Approximation* is of great importance both from the conceptual, and the computational point of view. In the next section, the Hartree-Fock equations are derived.

2.3.3   Hartree-Fock Approximation

In the Hartree-Fock (HF) approximation, the electronic wave function is given by a single Slater determinant. The spin orbitals $\{\varphi_i\}$ are then found by applying the variation principle in terms of one-electron atomic orbitals. Here, we outline this procedure without providing the details about mathematical operations. First, the wave function Ψ in equation (2.26) is substituted into the time-independent electronic Schrödinger equation (2.14), and the equation is solved for the electronic energy. The result, in atomic units, is given by

$$E = \sum_{i=1}^{N_e} \mathrm{h}_i + \frac{1}{2} \sum_{i=1}^{N_e} \sum_{j=1}^{N_e} (\mathrm{J}_{ij} - \mathrm{K}_{ij}) \qquad (2.35)$$

where h_i is the sum of the kinetic energy and the attraction energy for any electron that occupies orbital φ_i, and is given by

$$\mathrm{h}_i = \int \varphi_i(\mathbf{x}_1) \left(-\nabla_i^2(\mathbf{x}_1) - \sum_{I=1} \frac{Z_I}{|\mathbf{R}_I - \mathbf{r}_1|} \right) \varphi_i(\mathbf{x}_1) d\mathbf{x}_1 \qquad (2.36)$$

The notation $\mathbf{x}_1$ stands for any electron that is occupying orbital $\varphi_i(\mathbf{x}_1)$. That is to emphasize the formulation is focused on the orbitals, and not on the electrons. It is not important which electron has occupied φ_i. h_i is called the one-electron integral because it depends only on the integration of one orbital. In contrast, J_{ij} and K_{ij} depend on two orbitals, and are called two-electron integrals

$$\begin{aligned} \mathrm{J}_{ij} &= \int \int \varphi_i(\mathbf{x}_1)\varphi_j(\mathbf{x}_2)\frac{1}{|\mathbf{r}_2-\mathbf{r}_1|}\varphi_i(\mathbf{x}_1)\varphi_j(\mathbf{x}_2)d\mathbf{x}_1 d\mathbf{x}_2 \\ \mathrm{K}_{ij} &= \int \int \varphi_i(\mathbf{x}_1)\varphi_j(\mathbf{x}_2)\frac{1}{|\mathbf{r}_2-\mathbf{r}_1|}\varphi_i(\mathbf{x}_2)\varphi_j(\mathbf{x}_1)d\mathbf{x}_1 d\mathbf{x}_2 \end{aligned} \qquad (2.37)$$

J_{ij} is the Coulomb interaction between two orbitals φ_i and φ_j, and K_{ij} is the exchange energy. Corresponding to h_i, J_{ij}, and K_{ij}, three operators $\hat{\mathrm{h}}$, $\hat{\mathrm{J}}_j$, and $\hat{\mathrm{K}}_j$ are defined through their action on one-electron orbitals

$$\int \varphi_i(\mathbf{x}_1) \, \hat{h}(\mathbf{x}_1) \, \varphi_i(\mathbf{x}_1) \, dx_1 = h_i$$

$$\int \varphi_i(\mathbf{x}_1) \, \hat{J}_j(\mathbf{x}_1) \, \varphi_i(\mathbf{x}_1) \, dx_1 = J_{ij} \tag{2.38}$$

$$\int \varphi_i(\mathbf{x}_1) \, \hat{K}_j(\mathbf{x}_1) \, \varphi_i(\mathbf{x}_1) \, dx_1 = K_{ij}$$

The Hartree-Fock energy E_{HF} is obtained by applying the variation principle (2.28) to the energy equation (2.35). However, the orthonormality of the orbitals, equation (2.39), has to be satisfied at the same time.

$$\int \varphi_i(\mathbf{x}_1)\varphi_j(\mathbf{x}_1)dx_1 = \delta_{ij} \quad \forall i, j \tag{2.39}$$

This is done by augmenting the expression of energy by adding constraints (2.39) to equation (2.35)

$$\mathcal{E} = \sum_{i=1}^{N_e} h_i + \frac{1}{2} \sum_{i=1}^{N_e} \sum_{j=1}^{N_e} (J_{ij} - K_{ij}) - \sum_{i=1}^{N_e} \sum_{j=1}^{N_e} \varepsilon_{ij} \left[\int \varphi_i(\mathbf{x}_1)\varphi_j(\mathbf{x}_1)dx_1 - \delta_{ij} \right] \tag{2.40}$$

where $\{\varepsilon_{ij}\}$ are the Lagrange multipliers. Now, the variation of (2.40) with respect to any variation in orbitals $\{\varphi_i\}$ should vanish. The final result is a set of n similar equations for each orbital φ_i in the form

$$\left(\hat{h}(\mathbf{x}_1) + \sum_{j=1}^{N_e} \left[\hat{J}_j(\mathbf{x}_1) - \hat{K}_j(\mathbf{x}_1) \right] \right) \varphi_i(\mathbf{x}_1) = \sum_{j=1}^{N_e} \varepsilon_{ij}\phi_j(\mathbf{x}_1) \quad \forall i \tag{2.41}$$

By a proper transformation of orbitals, one can make matrix $\boldsymbol{\varepsilon}$ diagonal, *i.e.* $\varepsilon_{ij} = \varepsilon_j\delta_{ij}$, which yields the Hartree-Fock equations

$$\left(\hat{h}(\mathbf{x}_1) + \sum_{j=1}^{N_e} \left[\hat{J}_j(\mathbf{x}_1) - \hat{K}_j(\mathbf{x}_1) \right] \right) \varphi_i(\mathbf{x}_1) = \varepsilon_i\varphi_i(\mathbf{x}_1) \quad \forall i \tag{2.42}$$

The operator in parentheses is called the *Fock* operator. Introducing this in symbolic notation shows that the Hartree-Fock equations (2.42) are a system of eigenvalue equations

$$\hat{\mathbf{F}}(\mathbf{x}_1)\varphi_i(\mathbf{x}_1) = \varepsilon_i\varphi_i(\mathbf{x}_1) \qquad \forall i \tag{2.43}$$

The significance of equation (2.43) is that it has transformed the multi-electron equation (2.14) into a set of one-electron orbital equations. Nevertheless, these equations are not totally independent, but are coupled through the Fock operator $\hat{\mathbf{F}}$. The general procedure is to solve them simultaneously by a suitable iterative scheme until the solution orbitals satisfy all the equations in a consistent way.

Now, consider $\{\varepsilon_i\}$: if both sides of (2.43) are multiplied by $\varphi_i(\mathbf{x}_1)$ and integrated, one obtains

$$
\begin{aligned}
\varepsilon_i &= \int \varphi_i(\mathbf{x}_1)\,\hat{\mathbf{F}}\,\varphi_i(\mathbf{x}_1) \\
&= \mathrm{h}_i + \sum_{j=1}^{N_e}[\mathrm{J}_j - \mathrm{K}_j]
\end{aligned}
\tag{2.44}
$$

This means that ε_i is the energy of an electron when occupying orbital φ_i. This is a very useful picture of energy distribution in terms of one-electron orbitals that matches well with the concepts of the Molecular Orbital Model. The total electronic Hartree-Fock energy E_{HF} is not equal to the algebraic sum of $\{\varepsilon_i\}$, because the latter value counts the interaction energies twice. Thus the relationship between E_{HF} and the sum of orbital energies is

$$E_{\mathrm{HF}} = \sum_{i=1}^{N_e}\varepsilon_i - \frac{1}{2}\sum_{i=1}^{N_e}\sum_{j=1}^{N_e}[\mathrm{J}_{ij} - \mathrm{K}_{ij}] \tag{2.45}$$

The orbital energies (2.44) and E_{HF} in (2.45) have been developed based on a general set of spin-orbitals $\{\varphi_i\}$. The formulation is called the *unrestricted* Hartree-Fock (UHF), which uses different spatial functions for α and β spins. This is the most general case that can be applied to any chemical system. On the other hand, in the *restricted* Hartree-Fock (RHF) formulation, each orbital is constrained to have the same spatial functions for α and β. It requires that an even number of electron be present in the system such that each orbital becomes doubly occupied. It can be shown that orbital energies $\{\varepsilon_{i,\text{RHF}}\}$ can be obtained from

$$\varepsilon_{i,\text{RHF}} = h_i + \sum_{j=1}^{N_e/2} [2J_j - K_j] \tag{2.46}$$

and the RHF energy E_{RHF} is equal to

$$\begin{aligned} E_{\text{RHF}} &= 2 \sum_{i=1}^{N_e/2} h_i + \sum_{i=1}^{N_e/2} \sum_{j=1}^{N_e/2} [2J_{ij} - K_{ij}] \\ &= \sum_{i=1}^{N_e/2} [h_i + \varepsilon_i] \end{aligned} \tag{2.47}$$

In UHF, each orbital is characterized by the one-electron energy ε_i with either α (up) or β (down) spin configuration, while in RHF, each orbital is occupied by two electrons, one having α and the other β spin, and both electrons are at the same level of energy ε_i. The computations of the RHF approximation, when it is applicable, are considerably less demanding than the UHF approximation. The RHF energy is usually the starting point of calculations, unless it is not the lowest value. Finally, the total energy of the system; excluding rotational, translational, and internal vibrational components; is equal to the sum of its electronic energy E_{HF} and the inter-nuclear repulsions

$$E = E_{\text{HF}} + \sum_{I=1}^{N_n} \sum_{J>I}^{N_n} \frac{Z_I Z_J}{|\mathbf{R}_I - \mathbf{R}_J|} \tag{2.48}$$

So far, all the equations were derived in terms of molecular orbitals $\{\varphi_i\}$, whose explicit dependence on space coordinates has to be addressed for computational purposes. The usual procedure is to expand each φ_i as a linear combination of atomic orbitals (LCAO). In the next section, it is shown that the Hartree-Fock equations (2.43) lead to an eigenvalue problem, whose solution gives the coefficients of atomic orbitals in the molecular orbital expansions.

2.3.4 Roothaan Equations

Consider basis set $\{\chi_a(\mathbf{x}_1),\ a = 1..N_a\}$, whose elements are explicit functions of space coordinates, and implicit functions of spin. As discussed previously, the spin functions need to be present only from a mathematical point of view. We can expand the molecular orbitals $\{\varphi_i\}$ as a linear combination of $\{\chi_a\}$

$$\varphi_i(\mathbf{x}_1) = \sum_{a=1}^{N_a} \chi_a(\mathbf{x}_1)\, c_{ai} \tag{2.49}$$

where c_{ai} is the coefficient of basis function χ_a in the expansion of molecular orbital φ_i. Obviously, the more basis functions are used in the expansions, the more accurate will be the estimation of the orbital energies and the total energy in equation (2.48). However, due to computational constraints, one is usually limited to using only a finite number of basis functions. If equation (2.49) is substituted into the Hartree-Fock equations (2.43), it gives

$$\hat{\mathbf{F}}(\mathbf{x}_1) \sum_{a=1}^{N_a} \chi_a(\mathbf{x}_1)\, c_{ai} = \varepsilon_i \sum_{a=1}^{N_a} \chi_a(\mathbf{x}_1)\, c_{ai} \tag{2.50}$$

Multiplying both sides by $\chi_b(\mathbf{x}_1)$ and integrating over the range of $\mathbf{x}_1$ leads to

$$\sum_{a=1}^{N_a} c_{ai} \int \chi_b(\mathbf{x}_1)\, \hat{\mathbf{F}}(\mathbf{x}_1)\, \chi_a(\mathbf{x}_1)\, d\mathbf{x}_1 = \varepsilon_i \sum_{a=1}^{N_a} c_{ai} \int \chi_b(\mathbf{x}_1)\, \chi_a(\mathbf{x}_1)\, d\mathbf{x}_1 \tag{2.51}$$

Presumably, the two integrands are known functions of $\mathbf{x}_1$ and can be evaluated, while c_{ai} are to be determined. Denoting the left hand integral by F_{ba}, and the right hand integral by S_{ba}, equation (2.51) takes the following form

$$\sum_{a=1}^{N_a} F_{ba}\, c_{ai} = \varepsilon_i \sum_{a=1}^{N_a} S_{ba}\, c_{ai} \qquad \forall i \tag{2.52}$$

or in matrix form

$$\mathbf{F}\,\mathbf{c} = \mathbf{S}\,\mathbf{c}\,\boldsymbol{\varepsilon} \tag{2.53}$$

These are the Roothaan equations that are solved by quantum software, such as Gaussian, in the Hartree-Fock and hybrid methods. Although in equation (2.53), the coefficient matrix $\mathbf{c}$, and the eigenvalue vector $\boldsymbol{\varepsilon}$ are the only unknowns, they have to be found self-consistently. The Fock matrix $\mathbf{F}$, by definition, depends on the molecular orbitals $\{\varphi_i\}$, whose expansions in terms of the basis set $\{\chi_a\}$ depend on the coefficients c_{ai}. In practice, the solution procedure starts with an initial guess for the molecular orbitals, and (2.53) is solved for $\mathbf{c}$, which gives a new set of molecular orbitals. The loop is iterated until the difference between the input and the output molecular orbitals is negligible, thereby a so-called self-consistent field (SCF) solution has been obtained.

The expansion of the electronic wave function by Slater determinants and the Hartree-Fock approximation constitute the heart of all Molecular Orbital (or Wave Function) methods to solve the time-independent Schrödinger equation. Unfortunately, for many practical applications, the total energy of the system lacks desirable accuracy if only a single determinant wave function is used as in the HF approximation. On the other hand, an exact result for any given basis set is obtained if the expansion (2.27) includes all possible configuration interaction (CI) determinants. However, the full CI method is computationally too expensive to be feasible, and even if feasible for some cases it would be very time-consuming. Another category of methods is based on the charge density instead of wave function. The methodology is embedded in the theorem of the *Density Functional Theory* (DFT), and its further formulations. The DFT methodology was originally developed for the ground state of systems, but it has been extended to spin-polarized systems, excited states, and time dependent potentials as well. Both Wave Function and DFT methods indeed have the potential to accurately model any chemical system.

2.4   Density Functional Theory

Density Functional Theory (DFT) is based on the idea to replace the N-electron wave function Ψ as the solution of the Schrödinger equation with a scalar function ρ. The replacement means the energy, electronic configuration, and all other physical properties of the system are uniquely determined when ρ is known. While Ψ depends on the coordinates of N separate objects, ρ is a function of only three space coordinates $\mathbf{r} = (x, y, z)$,

$$\rho = \rho(\mathbf{r}) \tag{2.54}$$

or more generally four space-spin coordinates $\mathbf{x} = (x, y, z, s)$ such that

$$\rho = \rho(\mathbf{x}) \tag{2.55}$$

ρ, formally named *one-electron probability density*, equals the probability of finding any of the N electrons within the volume $d\mathbf{r}$ and having spin s when the remaining (N-1) electrons have positions and spins that are specified by Ψ [54]

$$\rho(\mathbf{x}) \equiv N \cdot \int |\Psi(\mathbf{x}, \mathbf{x_2}, \mathbf{x_3}, ..., \mathbf{x_N})|^2 \ d\mathbf{x_2} d\mathbf{x_3} ... d\mathbf{x_N} \tag{2.56}$$

or for a spin-free wave function and density $\rho(\mathbf{r})$ is defined as

$$\rho(\mathbf{r}) \equiv N \cdot \int |\Psi(\mathbf{r}, \mathbf{r_2}, \mathbf{r_3}, ..., \mathbf{r_N})|^2 \ d\mathbf{r_2} d\mathbf{r_3} ... d\mathbf{r_N} \tag{2.57}$$

The pre-factor N is to account for the electrons being indistinguishable from each other. ρ is subject to the constraint that its integral should sum up exactly to the number of electrons

$$\int \rho(\mathbf{x}) \, d\mathbf{x} = N$$

$$\text{or} \tag{2.58}$$

$$\int \rho(\mathbf{r}) \, d\mathbf{r} = N$$

This constraint results from the normality condition of the electronic wave function in equation (2.2).

The advantages of working with ρ instead of Ψ arise from the fact that the wave function of real systems is typically a complicated and complex function of $3N$ or $4N$ electronic coordinates, and not amenable to experimental measurements. On the other hand, the electronic density is a physically measurable quantity that remains a function of only 4 variables independent of the system size.

Initially, the idea of using the electronic density was proposed by Thomas and Fermi in 1927-28, who expressed the electronic kinetic energy of a single atom as a functional of ρ. They estimated the total energy as a function of only the probability density. The functional lacked the exchange and correlation effects, thus even with subsequent improvements, the Thomas-Fermi model remained inaccurate for most systems. The exchange energy refers to the interaction of two overlapping electronic wave functions, and the correlation energy accounts for the inter-dependency of the two wave functions. In the Hartree-Fock approximation, the exchange energy of electrons with parallel spin is captured, while that of the anti-parallel spins is neglected. It was not until 1964 and 1965, when two papers by Hohenberg-Kohn and Kohn-Sham paved the way to the practical applications of density in the DFT method.

2.4.1 Hohenberg-Kohn Theorems

In 1964, Hohenberg and Kohn [50] published their pioneering work, in which they proved the existence and uniqueness of an electronic density that corresponded to the ground state of a system. The following discussions and formulations are developed based on the material in their work. First, a system with an arbitrary number of electrons is considered to be influenced by an external potential $v_{ext}(\mathbf{r})$, principally given by Coulombic attractions and repulsions. The Hamiltonian of the system is written as the sum of three contributions

$$H = T_{kin} + V_{ext} + U_{ee} \tag{2.59}$$

where the energy terms in atomic units are defined as

$$
\begin{aligned}
T_{\text{kin}} &\equiv \frac{1}{2} \int \nabla \Psi^*(\mathbf{r}) \, \nabla \Psi(\mathbf{r}) \, d\mathbf{r} \\
V_{\text{ext}} &\equiv \int \nabla \Psi^*(\mathbf{r}) \, v_{\text{ext}}(\mathbf{r}) \, \nabla \Psi(\mathbf{r}) \, d\mathbf{r} \\
U_{\text{ee}} &\equiv \frac{1}{2} \int \Psi^*(\mathbf{r}) \, \Psi^*(\mathbf{r}') \, \frac{1}{|\mathbf{r} - \mathbf{r}'|} \, \Psi(\mathbf{r}') \, \Psi(\mathbf{r}) \, d\mathbf{r}' d\mathbf{r}
\end{aligned}
\tag{2.60}
$$

The wave function solution of the Schrödinger equation and the electron density ρ will certainly depend on the external potential $v_{\text{ext}}(\mathbf{r})$, which is primarily the Coulombic attraction of nuclei. Assuming a non-degenerate ground state, it is proved that $v_{\text{ext}}(\mathbf{r})$ is in turn a unique functional of ρ within an additive constant. Otherwise, having two different external potentials $v_{\text{ext}}(\mathbf{r})$ and $v'_{\text{ext}}(\mathbf{r})$ would lead to a contradiction. This constant is of no special significance, and is arbitrary as the origin of energy is.

Consider two potentials $v_{\text{ext}}(\mathbf{r})$ and $v'_{\text{ext}}(\mathbf{r})$ that correspond to two different wave functions Ψ and Ψ', and two different Hamiltonian H and H', but both follow from the same electronic density $\rho(\mathbf{r})$. The solution of the Schrödinger equation for $v_{\text{ext}}(\mathbf{r})$ and $v'_{\text{ext}}(\mathbf{r})$ gives rise to two ground state energies E and E'. If instead of Ψ, Ψ' is used to estimate the energy of the system, according to the variation principle, a higher energy will be obtained, because Ψ' is not the eigenfunction of H. Thus

$$
E = \int \Psi^* \hat{H} \, \Psi \, d\mathbf{r} \;\; < \int \Psi'^* \hat{H} \, \Psi' \, d\mathbf{r}
\tag{2.61}
$$

Expanding the right hand side of the inequality (2.61) yields

$$
\int \Psi'^* \hat{H} \, \Psi' \, d\mathbf{r} = \int \Psi'^* \left[\hat{T}_{\text{kin}} + \hat{V}_{\text{ext}} + \hat{U}_{\text{ee}} \right] \Psi' \, d\mathbf{r}
\tag{2.62}
$$

where $(\hat{\cdot})$ represents the quantum operator corresponding to equations (2.59) and (2.60). Noting that $\hat{T}_{\text{kin}}$ and $\hat{U}_{\text{ee}}$ are the same for H and H', introduce $\hat{H}'$ in (2.62)

$$
\begin{aligned}
\int \Psi'^* \, \hat{H} \, \Psi' \, d\mathbf{r} &= \int \Psi'^* \left[\hat{T}_{\text{kin}} + \hat{V}_{\text{ext}} + \hat{U}_{\text{ee}} \right] \Psi' \, d\mathbf{r} \\
&= \int \Psi'^* \left[\hat{T}_{\text{kin}} + \hat{V}'_{\text{ext}} + \hat{U}_{\text{ee}} + \hat{V}_{\text{ext}} - \hat{V}'_{\text{ext}} \right] \Psi' \, d\mathbf{r} \\
&= \int \Psi'^* \left[\hat{H}' + \hat{V}_{\text{ext}} - \hat{V}'_{\text{ext}} \right] \Psi' \, d\mathbf{r} \qquad (2.63) \\
&= \int \Psi'^* \, \hat{H}' \, \Psi' \, d\mathbf{r} + \int \left[v_{\text{ext}}(\mathbf{r}) - v'_{\text{ext}}(\mathbf{r}) \right] \rho(\mathbf{r}) \, d\mathbf{r} \\
&= E' + \int \left[v_{\text{ext}}(\mathbf{r}) - v'_{\text{ext}}(\mathbf{r}) \right] \rho(\mathbf{r}) \, d\mathbf{r}
\end{aligned}
$$

Substitution of (2.63) in (2.61) gives

$$
E < E' + \int \left[v_{\text{ext}}(\mathbf{r}) - v'_{\text{ext}}(\mathbf{r}) \right] \rho(\mathbf{r}) \, d\mathbf{r} \qquad (2.64)
$$

Similarly, if one starts from Ψ and H', it would result to

$$
E' < E + \int \left[v'_{\text{ext}}(\mathbf{r}) - v_{\text{ext}}(\mathbf{r}) \right] \rho(\mathbf{r}) \, d\mathbf{r} \qquad (2.65)
$$

Adding (2.64) and (2.65) leads to contradiction

$$
E' + E < E + E' \qquad (2.66)
$$

Thus $v_{\text{ext}}(\mathbf{r})$, within a constant, is uniquely determined by density ρ. The system Hamiltonian, wave function, ground state energy, and all other physical properties now become functions of ρ. In other words, the electron density can replace Ψ, because it carries the same information about the system. For instance, the energy is expressed as a function of ρ

$$
E[\rho(\mathbf{r})] = T_{\text{kin}}[\rho(\mathbf{r})] + U_{\text{ee}}[\rho(\mathbf{r})] + V_{\text{ext}}[\rho(\mathbf{r})] \qquad (2.67)
$$

The first two terms on the right hand side, although contain the corresponding information, are not explicit functions of the external potential $v_{\text{ext}}(\mathbf{r})$. When grouped together, they make the Hohenberg-Kohn universal functional F_{HK} defined as

$$F_{HK}[\rho(\mathbf{r})] = T_{\text{kin}}[\rho(\mathbf{r})] + U_{\text{ee}}[\rho(\mathbf{r})] \qquad (2.68)$$

The form of F_{HK} is the same for all systems no matter how many electrons are present in the system and what form v_{ext} might have. This should not be a surprising conclusion, as it was shown that all the information about a system has already been reflected in the density itself. For example, the number of electrons can be obtained through equation (2.58).

The second theorem of Hohenberg and Kohn is the application of the variational principle to the energy as a function of density. It states that if a trial density $\tilde{\rho}(\mathbf{r})$ is used such that it satisfies constraint (2.58), it would give rise to energy $\tilde{E}$ which is higher than the ground state energy E

$$E[\rho(\mathbf{r})] < E[\tilde{\rho}(\mathbf{r})] = \tilde{E} \qquad (2.69)$$

The Hohenberg-Kohn theorems are no more than the proofs of the existence of a unique electron density which corresponds to the ground state energy. Although they are exact theorems, in the sense that no approximation is made in their proof, they do not provide any clue as what form the universal functional F_{HK} might have.

One year later in 1965, Kohn and Sham tried to answer this question by deriving approximate equations, which constitute the starting point of DFT-based computations. However, even after decades of the invention of DFT and the Kohn-Sham equations, the exact form of F_{HK} is still unknown.

2.4.2 Kohn-Sham Equations

In 1965, Kohn and Sham introduced the one-electron orbitals, similar to the Hartree-Fock (HF) approximation, to solve the many-electron problem of the Schrödinger equation within the DFT formulation developed through Hohenberg-Kohn theorems. In their formulation, the main contributions to the kinetic energy and electron-electron interactions are captured exactly by treating electrons as non-interacting bodies. The remaining parts then are collected in a single energy term, the exchange-correlation energy that, once known, determines the exact density and exact energy of the system. The Kohn-Sham equations transformed DFT from a set of purely theoretical theorems about existence and uniqueness into the most cost effective computational tool in Chemistry and Material Science.

Consider a set of one-electron orbitals $\{\psi_i\}$ as functions of $\mathbf{r}_i$. The expressions for the kinetic energy and the density can be written as

$$
\begin{aligned}
T_{\text{kin}} &= \sum_i n_i \int \psi_i^*(\mathbf{r}) \left(-\frac{\nabla_i^2}{2} \right) \psi_i(\mathbf{r}) \, d\mathbf{r} \\
\rho(\mathbf{r}) &= \sum_i n_i |\psi_i(\mathbf{r})|^2
\end{aligned}
\tag{2.70}
$$

where n_i is the occupation number of the ith orbital. n_i must lie between 0 and 1 inclusively, according to the Pauli principle. For a real system comprised of interacting electrons, expansions (2.70) become exact if an infinite number of terms and orbitals are included. Kohn and Sham suggested to formulate the problem in terms of only N fully occupied orbitals similar to the HF formulation, that is

$$
\begin{aligned}
T_{KS} &= \sum_i^N \int \psi_i^*(\mathbf{r}) \left(-\frac{\nabla_i^2}{2} \right) \psi_i(\mathbf{r}) \, d\mathbf{r} \\
\rho(\mathbf{r}) &= \sum_i^N |\psi_i(\mathbf{r})|^2
\end{aligned}
\tag{2.71}
$$

which are special cases of (2.70) with $n_i = 1$ for N orbitals and $n_i = 0$ for the rest. Here, the subscript KS is used to emphasize the difference between non-interacting kinetic energy and the interacting one in equations (2.70).

For T_{KS} to be the exact kinetic energy of N non-interacting electrons, orbitals $\{\psi_i\}$ must be orthonormal

$$\int \psi_i^*(\mathbf{r})\,\psi_j(\mathbf{r})\,d\mathbf{r} = \delta_{ij} \qquad \forall i, j \tag{2.72}$$

Equation (2.71) automatically satisfies constraint (2.58) on ρ, but now as a set of constraints on $\{\psi_i\}$. Next, the exchange-correlation energy E_{xc} is defined such that it contains the difference between T_{kin} and T_{KS}, and the difference between U_{ee} and its classical Coulomb's potential energy J

$$E_{\mathrm{xc}}[\rho] \equiv (T_{\mathrm{kin}}[\rho] - T_{KS}[\rho]) + (U_{\mathrm{ee}} - J[\rho]) \tag{2.73}$$

where J is given as a function of ρ by

$$J[\rho] \equiv \frac{1}{2} \int \int \frac{\rho(\mathbf{r})\,\rho(\mathbf{r}')}{|\mathbf{r}' - \mathbf{r}|}\,d\mathbf{r}'\,d\mathbf{r} \tag{2.74}$$

With these definitions, the Hohenberg-Kohn functional F_{HK} in equation (2.68) is re-written as

$$F_{HK} = T_{KS}[\rho] + J[\rho] + E_{\mathrm{xc}}[\rho] \tag{2.75}$$

and the DFT energy (2.67) is now equal to

$$E[\rho] = T_{KS}[\rho] + J[\rho] + E_{\mathrm{xc}}[\rho] + V_{\mathrm{ext}} \tag{2.76}$$

To arrive at the complete form of the problem, one can add constraints (2.72) by Lagrange multipliers, similar to the HF formulation (2.40), and define an augmented functional Υ

$$\Upsilon = E[\rho] - \sum_i^N \sum_j^N \varepsilon_{ij} \left[\int \psi_i^*(\mathbf{r}) \, \psi_j(\mathbf{r}) \, d\mathbf{r} - \delta_{ij} \right] \tag{2.77}$$

Finally, a Slater determinantal form as in equation (2.26) is assumed for the total electronic wave function Ψ

$$\Psi = \frac{1}{\sqrt{N!}} det(\psi_1, \psi_2, ..., \psi_N) \tag{2.78}$$

With this wave function, it can be shown that the time-independent Schrödinger equation for Ψ leads to an eigenvalue system in terms of one-electron orbitals $\{\psi_i\}$, similar to equations (2.43)

$$\left[-\frac{\nabla^2}{2} + v_{\text{eff}}(\mathbf{r}) \right] \psi_i = \varepsilon_i \psi_i \qquad \forall i \tag{2.79}$$

where the effective potential $v_{\text{eff}}(\mathbf{r})$ equals

$$v_{\text{eff}}(\mathbf{r}) = v_{\text{ext}}(\mathbf{r}) + \int \frac{\rho(\mathbf{r}')}{|\mathbf{r}' - \mathbf{r}|} \, d\mathbf{r}' + \frac{\delta E_{\text{xc}}[\rho]}{\delta \rho(\mathbf{r})} \tag{2.80}$$

The above equations, first derived by Kohn and Sham in 1965, need to be solved self-consistently. Usually one starts with an initial assumption for ρ, constructs v_{eff} from (2.80), solves equations (2.79), and finds a new estimation for ρ using (2.71). The procedure is repeated until convergence is achieved. Then the energy is evaluated from equation (2.76).

It has to be emphasized here that the Kohn-Sham equations above do not have any approximation within the Hohenberg-Kohn theorems. They can still lead to the exact total energy, if the exact form of E_{xc} is used. The major difference between the HF and DFT formulations is that HF is based on an approximate formulation, where the correlation effects cannot be incorporated as exactly as in DFT . On the other hand,

while there is no systematic way for improving the DFT energy, one can systematically improve the Wave-Function approximations by adding more Slater determinants
using the configuration interaction formulation. The other difference between these
two methods is that, unlike in HF, the DFT orbitals and their corresponding energies
are merely mathematical tools for solving the Kohn-Sham equation, and carry no
physical meaning.

2.4.3 Exchange-Correlation Functionals

The success of the DFT approach depends on the precision of the applied exchange-
correlation energy E_{xc}. For a homogeneous electron gas, as in metals, E_{xc} is a function
of the density value ρ only. For systems with a non-homogeneous electron gas, it
depends on the first and higher order gradients of ρ. Using only the value of the
density leads to the Local Density Approximation (LDA)

$$E_{\text{xc}}^{LDA}[\rho] = \int \rho(\mathbf{r})\, \varepsilon_{\text{xc}}(\rho)\, d\mathbf{r} \qquad (2.81)$$

ε_{xc} reflects the exchange-correlation energy per particle of a uniform gas of electrons
with density ρ. In LDA, the exchange-correlation potential in equation (2.80) takes
the form

$$\frac{\delta E_{\text{xc}}^{LDA}[\rho]}{\delta \rho(\mathbf{r})} = \varepsilon_{\text{xc}}(\rho) + \rho(\mathbf{r})\frac{\partial \varepsilon_{\text{xc}}(\rho)}{\partial \rho} \qquad (2.82)$$

$\varepsilon_{\text{xc}}(\rho)$ has been extensively studied in the literature, and several forms for its dependence on ρ have been suggested. It is often split further into exchange and correlation
contributions, and each part is studied separately.

$$\varepsilon_{\text{xc}}(\rho) = \varepsilon_{\text{x}}(\rho) + \varepsilon_{\text{c}}(\rho) \qquad (2.83)$$

The exchange part $\varepsilon_x(\rho)$, apart from a numerical coefficient, is equal to the exchange energy in the X_α formulation derived by Slater, given by

$$\varepsilon_x(\rho) = -\frac{3}{4}\sqrt[3]{\frac{3\,\rho(\mathbf{r})}{\pi}} \tag{2.84}$$

The LDA approach is formally applicable only to systems with a slowly varying density. Nevertheless, it has been surprisingly successful for some highly inhomogeneous systems. This unexpected success of LDA is usually attributed to the cancellation of errors. In particular, LDA overestimates the binding energies because of its poor performance in describing isolated atoms, which typically have highly inhomogeneous densities.

To incorporate spin in DFT, the electron density for an unrestricted electron density is written as the sum of two spin densities ρ_α and ρ_β

$$\rho(\mathbf{r}) = \rho_\alpha(\mathbf{r}) + \rho_\beta(\mathbf{r}) \tag{2.85}$$

Correspondingly, the LDA is extended to Local Spin Density (LSD) approximation

$$E_{xc}^{LSD}[\rho_\alpha, \rho_\beta] = \int \rho(\mathbf{r})\,\varepsilon_{xc}(\rho_\alpha, \rho_\beta)\,d\mathbf{r} \tag{2.86}$$

There have been theoretical improvements to LDA, such as the Generalized Gradient Approximation (GGA), and the Green function approach, called GW. In GGA, the exchange-correlation energy is a functional of not only the local density value, but also of its first gradient $\nabla\rho$

$$E_{xc}^{GGA}[\rho_\alpha, \rho_\beta] = \int f(\rho_\alpha, \rho_\beta, \nabla\rho_\alpha, \nabla\rho_\beta)\,d\mathbf{r} \tag{2.87}$$

E_{xc}^{GGA} is usually split into its exchange and correlation components, which are treated separately.

$$E_{xc}^{GGA} = E_x^{GGA} + E_c^{GGA} \tag{2.88}$$

Practically, expressions for E_x^{GGA} and E_c^{GGA} are developed such that the mean error of the results compared with tabulated data sets be minimized for a given model. Thus, in general, the developed expressions for E_{xc}^{GGA} do not provide any physical understanding in terms of the functionals.

The exchange part E_x^{GGA} is written as

$$E_x^{GGA} = E_x^{LSD} - \sum_{\sigma=\alpha,\beta} \int F(s_\sigma) \, \rho_\sigma^{4/3}(\mathbf{r}) \, d\mathbf{r} \qquad (2.89)$$

where s_σ is the reduced density gradient for spin σ, and is equal to

$$s_\sigma \equiv \frac{|\nabla \rho_\sigma(\mathbf{r})|}{\rho_\sigma^{4/3}(\mathbf{r})} \qquad \sigma = \alpha, \beta \qquad (2.90)$$

The 4/3 value for the power of ρ is to make s_σ a dimensionless quantity. s_σ indicates the inhomogeneity in the density. It takes on large values where the density is changing rapidly, or where the density is vanishing at distances far away from the nuclei. s_σ is small at regions where the density gradient is small, such as in bonding regions, or where the density is high as regions close to nuclei. In particular, for a homogeneous electron gas s_σ equals zero everywhere in space.

Several forms have been suggested for function $F(s_\sigma)$. Here two well-known forms are presented. The first form is a GGA-based function that was developed by Becke in 1988 [22], called B88, to asymptotically match the exchange energy far from a finite system. B88 contains one empirical parameter $\beta = 0.0042$, which has been fitted to give the least error for known exchange energies of the rare gases.

$$F^{B88}(s_\sigma) = \frac{\beta s_\sigma}{1 + 6\beta s_\sigma \, \sinh^{-1} s_\sigma} \qquad (2.91)$$

The other forms of F often take a rational function in terms of s_σ. As an example, Perdew's 1986 [86] (P86), given below, is a parameter-free function.

$$F^{P86}(s_\sigma) = \left(1 + 1.296 \, \bar{s}_\sigma^2 + 14 \, \bar{s}_\sigma^4 + 0.2 \, \bar{s}_\sigma^6\right)^{1/15} \qquad (2.92)$$

where $\bar{s}_\sigma$ is defined as

$$\bar{s}_\sigma \equiv \frac{s_\sigma}{\sqrt[3]{24\pi^2}} \tag{2.93}$$

Similarly, analytical expressions for the correlation energy have been proposed in the literature. One of the most accepted formula is that of Vosko, Wilk and Nusair [103], known as E_c^{VWN}. The formula for E_c^{VWN} is an interpolation based on the accurate numerical calculations of Ceperly and Alder [28], who used quantum Monte-Carlo method to estimate the correlation energy of a uniform gas.

We now turn to the hybrid methods, and in particular to B3LYP, which is one of the most widely used computational methods in the literature, and in this book too. Hybrid methods are based on the DFT approach, but estimate the exchange-correlation energy by mixing DFT functionals with those of the Wave Function method. The mixing is controlled by one or more constant parameters. For example, B3LYP [96] uses the three-parameter formula of Becke [22], and the correlation functional of Lee-Yang-Parr [56]. E_{xc}^{B3LYP} is given explicitly as [96]

$$E_{xc}^{B3LYP} = (1 - a_0)\, E_x^{LSD} + a_0\, E_x^{HF} + a_x\, \Delta E_x^{B88} + a_c\, E_c^{LYP} + (1 - a_c)\, E_c^{VWN} \tag{2.94}$$

where
E_x^{LSD} is the standard local exchange energy,
E_x^{HF} is the exact exchange energy in the Hartree-Fock method,
ΔE_x^{B88} is Becke's gradient correction to the exchange functional,
E_c^{LYP} and E_c^{VWN} are the correlation energy of Lee-Yang-Parr, and Vosko-Wilk-Nusair, respectively. Based on fitting the heats of formation of small molecules, Becke proposed $a_0 = 0.2$, $a_x = 0.72$, and $a_c = 0.81$.

There are also other hybrid methods with different flavors and modifications. Several researchers have studied the accuracy of DFT hybrids, and have concluded that each flavor gives results more or less comparable to the other hybrids. Since discussing other formulations serves more to a theoretical point of view, which is not the focus of this book, no further detail on the computational theories is presented here. Instead, we switch to the other side of the computational coin, the theory behind atomic basis sets. Basis sets are mathematical functions, or better said mathematical tools, based on which the wave function solution is built. The trial wave function is then optimized within a given theoretical approach such as Hartree-Fock or DFT by changing the expansion coefficients.

2.5   Basis Sets

Any trial wave function solution for the time-independent Schrödinger equation (2.11) can be written in the form of either Slater determinants, or a sum over *plane wave* functions. A plane wave is the mathematical solution of the wave equation for a free particle, *i.e.*, a particle which experiences no potential. In general, it is a function of both space $\mathbf{r}$ and time t, given by

$$u(\mathbf{r}, t) = a \, \exp\left[i\left(\mathbf{k}.\mathbf{r} - \omega \, t\right)\right] \tag{2.95}$$

where a is the amplitude, $\mathbf{k}$ is the wave vector, and ω is the angular frequency. Dropping the time dependency, a candidate wave function $\Psi(\mathbf{r})$ can be expressed as a linear combination of plane waves

$$\Psi(\mathbf{r}) = \sum_{\mathbf{k}}^{N_{\mathbf{k}}} C_{\mathbf{k}} \, \exp\left(i\mathbf{k}.\mathbf{r}\right) \tag{2.96}$$

Coefficients $C_{\mathbf{k}}$ are to be determined through the variational principle. The plane wave approach is the natural way of dealing with periodic systems, where the Bloch theorem applies. The Bloch theorem states that the eigen functions of a periodic system, except for a phase factor, are the same for all equivalent cells. Therefore, the theorem can be used to reduce the problem of large systems to the computation of only one periodic cell. This is most suitable in solid state and condensed matter systems, where periodicity of structures and boundary conditions plays a major role in determining the chemical properties. The energy levels in such systems are often studied in terms of energy bands, which correspond to the set of wave vectors $\{\mathbf{k}\}$ in equation (2.96).

Compared to the basis sets that use atomic orbitals, and which are discussed below, the plane waves are delocalized functions that do not belong to any specific atomic center, but rather are spread over space. The plane wave basis sets cannot represent local properties, such as bonding in gas phase, in which localized functions are more required than non-local functions. Consequently, to attain a certain accuracy, one often needs more basis functions in the plane wave approach than when using atomic orbitals. We now focus on atomic orbitals as they have been widely used in this book for quantum computations.

Similar to equation (2.95), the molecular orbitals (MO) being the building blocks of Slater determinants (2.26), can be written as linear combinations of atomic orbitals $\{\chi_j\}$

$$\phi_i = \sum_{j=1} C_{ij} \chi_j \qquad (2.97)$$

Atomic orbitals (AO) are localized functions in terms of the coordinates of their centers. Orbital centers are often chosen to be the nuclei of atoms. The overall shape and orientation of any set of AO in a molecule are obtained by the numerical or analytical solution of the Schrödinger equation. The shape of an orbital refers to its radial, while the orientation refers to its angular distribution over space.

Analytical solutions for the Schrödinger equation are only known for hydrogen atom, and are called hydrogen orbitals. For other atoms, the basis sets are built in the same form as for Hydrogen, but with different parameters that are determined numerically. The Hydrogen-like orbitals, called Slater-type-orbitals (STO), are expressed in terms of atom-centered polar coordinates (r, θ, ϕ) [31]

$$\chi^{STO}(r, \theta, \phi; \zeta, n, l, m) = Y_l^m(\theta, \phi) \frac{(2\zeta)^{n+1/2}}{\sqrt{(2n)!}} \, r^{n-1} \, \exp(-\zeta r) \qquad (2.98)$$

where n, l, m are, respectively, the principle and angular momentum quantum numbers. ζ is the exponent that controls the diffusivity of the radial component, and Y_l^m is the spherical harmonic function that determines the angular orientation of the orbital. The fractional part in (2.98) is to ensure that the orbital is normalized.

Slater orbitals have the correct exponential decay as $r \to \infty$. Also the 1s-STO $(n + l + m = 0)$ has the correct cusp at $r = 0$. Despite the correct behavior, there is no analytical formula to evaluate the two-electron integrals in equation (2.37) when STOs are used. Instead, these integrals have to be estimated numerically. Because the number of these integrals increases with the 4th power of the number of basis functions, numerical evaluation of STO integrals to a desired accuracy can be very demanding from a computational point of view. The usage of Slater orbitals, although still present, is now very limited, and almost all quantum programs use Gaussian-type orbitals (GTO). The general form of a GTO in atomic-centered Cartesian coordinates (x, y, z) is given by [31]

$$\chi^{GTO}(x, y, z; \alpha, i, j, k) = Y_l^m(\theta, \phi) \, N \, x^i \, y^j \, z^k \, \exp\left[-\alpha \left(x^2 + y^2 + z^2\right)\right] \qquad (2.99)$$

with

$$N = \left(\frac{2\alpha}{\pi}\right)^{3/4} \left[\frac{(2\alpha)^{i+j+k} \, i! \, j! \, k!}{(2i)! \, (2j)! \, (2k)!}\right]^{1/2} \qquad (2.100)$$

where α controls the width of the orbital, similar to ζ in equation (2.98). i, j, and k are positive integers indicating the nature of the orbital. If $i = j = k$, the orbital has spherical symmetry, and is a s-type GTO. When only one of these three indices is unity, the function has axial symmetry along one Cartesian direction, and thus the orbital is a p-type GTO. Therefore, there are three p-type orbitals: p_x, p_y, and

p_z corresponding to three Cartesian axes. When summation $i + j + k$ equals two, a d-type GTO is obtained, leading to six possible Cartesian pre-factors: x^2, y^2, z^2, xy, xz and yz. However, only five functions are required to span all eigen functions of the Schrödinger equation for the z-component of the angular momentum for d-type orbitals. These five functions are usually chosen as xy, xz, yz, $x^2 - y^2$ and $3z^2 - r^2$. The first three of this set are common with the initial six d- functions, while the last two can be obtained as linear combinations of the square terms. The sixth function $r^2 = x^2 + y^2 + z^2$ has spherical symmetry, and therefore is a s-type function. From all these six Cartesian functions, the first five are called *canonical*. As the indexing increases, more Cartesian functions remain as non-canonical. For example, there are 10 Cartesian and 7 canonical functions of the f-type. A quantum program may choose all the Cartesian functions, or only the canonical forms in the implementation of the basis sets.

In contrast to Slater orbitals (2.98), Gaussian orbitals (2.99) do not have the correct behavior neither at $r = 0$ nor when $r \to \infty$. Figure 2.1 compares the radial component of $1s$-STO with a $1s$-GTO. Instead of a cusp, the Gaussian orbital has a zero slope at the origin, and falls off more rapidly with increasing r. On the other hand, Gaussian functions give rise to analytical evaluation of the electronic integrals, which can be found only numerically for Slater orbitals. To illustrate this, consider two $1s$ Gaussian orbitals $\chi_A(\alpha, \mathbf{r} - \mathbf{R}_A)$ and $\chi_B(\beta, \mathbf{r} - \mathbf{R}_B)$ that are centered on two atoms A and B. α and β are the exponents of the radial function, while $\mathbf{R}_A$ and $\mathbf{R}_B$ denote the locations of A and B, respectively. It can be shown that the product of these two orbitals is another Gaussian function $\chi_C(\gamma, \mathbf{r} - \mathbf{R}_C)$

$$\chi_A(\alpha, \mathbf{r} - \mathbf{R}_A) \cdot \chi_B(\beta, \mathbf{r} - \mathbf{R}_B) = K_{AB}\, \chi_C(\gamma, \mathbf{r} - \mathbf{R}_C) \qquad (2.101)$$

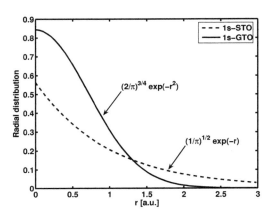

Figure 2.1: Comparison of the radial component of $1s$-STO with a $1s$-GTO

whose amplitude, exponent, and center are given by

$$K_{AB} = \left(\frac{2\alpha\beta}{\pi(\alpha+\beta)}\right)^{3/4} \exp\left(-\frac{\alpha\beta}{\alpha+\beta}|\mathbf{R}_A - \mathbf{R}_B|^2\right)$$
$$\gamma = \alpha + \beta \qquad\qquad\qquad (2.102)$$
$$\mathbf{R}_C = \frac{\alpha\,\mathbf{R}_A + \beta\,\mathbf{R}_B}{\alpha+\beta}$$

As an application of (2.101), let us show that the overlap integral between $\chi_A(\alpha, \mathbf{r} - \mathbf{R}_A)$ and $\chi_B(\beta, \mathbf{r} - \mathbf{R}_B)$ can be evaluated analytically.

$$\int \chi_A(\alpha, \mathbf{r} - \mathbf{R}_A) \cdot \chi_B(\beta, \mathbf{r} - \mathbf{R}_B) d\mathbf{r} = K_{AB} \int \chi_C(\gamma, \mathbf{r} - \mathbf{R}_C) d\mathbf{r}$$
$$= K_{AB} \cdot 4\pi \int_0^\infty r^2 \exp\left(-r^2\right) dr \qquad (2.103)$$
$$= K_{AB} \left(\frac{\pi}{\gamma}\right)^{3/2}$$

Similar analytical expressions can be obtained for other integrals, such as the two-electron and kinetic energy integrals. Thus Gaussian orbitals present attractive computational benefits over Slater orbitals, but lack the correct functional behavior. One way around this situation is to linearly combine a few GTOs such that computations can still benefit from analytical expressions without losing much of accuracy. The orbitals so-obtained, called *contracted* Gaussian orbitals (CGOs), can be written as

$$\chi^{CGO}(\mathbf{r} - \mathbf{R}_A) = \sum_{p=1}^{L} d_p\, \chi^{GTO}(\alpha_p, \mathbf{r} - \mathbf{R}_A) \tag{2.104}$$

where L is the contraction length, d_p is the contraction coefficient, and χ^{GTO} are now called the *primitive Gaussians*. The idea behind contraction is to use a few Gaussian primitives with fixed exponents and coefficients such that they best approximate a target function such as a Slater orbital. Fitted to Slater functions, the contracted orbitals are called STO-LG, with L denoting the length of the contraction, that is the number of Gaussian primitives. Suppose one wants to approximate a 1s-STO with three successive contraction length $L = 1, 2, 3$ as in (2.104). In each case, we need to determine the set of d_p and α_p in order to maximize overlap between resultant function χ_{1s}^{CGO} and $\chi_{1s}^{STO}(\zeta = 1.0)$. This is an optimization problem in integral form: one needs to maximize the integral

$$
\begin{aligned}
S &= \int \chi_{1s}^{STO}(\zeta = 1.0) \cdot \chi_{1s}^{CGO}\, d\mathbf{r} \\
&= \int \chi_{1s}^{STO}(\zeta = 1.0) \cdot \sum_{p=1}^{L} d_p\, \chi^{GTO}(\alpha_p, \mathbf{r} - \mathbf{R}_A)\, d\mathbf{r} \qquad L = 1, 2, 3
\end{aligned}
\tag{2.105}
$$

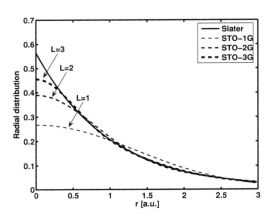

Figure 2.2: Comparison of the radial component of $1s$-STO with contracted STO-LG, $L = 1, 2, 3$ orbitals.

Numerical solution of problem (2.105) gives

$$L = 1 : \quad \left\{ \; d_1 = 1.0 \qquad , \; \alpha_1 = 0.270950 \right.$$

$$L = 2 : \quad \left\{ \begin{array}{ll} d_1 = 0.678914, & \alpha_1 = 0.151623 \\ d_2 = 0.430129, & \alpha_2 = 0.851819 \end{array} \right.$$

$$L = 3 : \quad \left\{ \begin{array}{ll} d_1 = 0.444464, & \alpha_1 = 0.109818 \\ d_2 = 0.535328, & \alpha_2 = 0.405771 \\ d_3 = 0.154329, & \alpha_3 = 2.22766 \end{array} \right.$$

Figure 2.2 compares three sets of contracted GTO with $1s$-STO. As expected, with increasing the number of contraction functions, a better fit to the target function is obtained. Although using a large value for L leads to a better approximation of

Slater-type functions, the corresponding computational cost is not as cheap. As mentioned before, the number of two-electron integrals scales with the 4th power of the number of the involved basis functions, thus an increase in the number of integrals may cancel the benefit of using analytical formulae. As a rule of thumb, the contraction of 3 or 4 primitives Gaussians is usually a cost-effective choice.

STO-3G is often referred to as the *minimal* or *single-ζ* basis set, because it uses for every atomic orbital only one exponent ζ corresponding to one Slater function. Aside from qualitative results, the minimal basis set can hardly lead to quantitatively accurate results. More flexibility and thus accuracy can be gained if one allows two or more Slater functions be used for each atomic orbital. Such improved basis sets are called *double-* , *triple-*,..., *multiple-ζ* functions. The more ζ exponents are used in a basis set, the more flexible will it be in the molecular environment.

Another point is that most chemical activities are processes which occur between the outermost (valence) orbitals, while the inner (core) orbitals remain almost untouched. Given a fixed number of total basis functions, better accuracy can be obtained if one uses more flexible functions for the valence orbitals, and treats the core electrons by a minimal set. Treating the core and valence orbitals differently is called the *split-valence* approach. Most recent basis sets such as 3-21G or 6-31G are split-valence functions. In this naming convention, the first number denotes the number of primitive Gaussians in the core orbitals, while the rest determine the number of exponents and their corresponding primitives. In set 6-31G, every core orbital uses 6 primitive Gaussian functions, and each valence orbital is a double-ζ that has 3 primitives for the first and 1 for the second function.

By construct, basis functions are optimized according to the orbitals of isolated atoms. More than often however, the atomic orbitals are distorted (*polarized*) when they are exposed to the field of other atoms in a molecule. The polarization effect can be incorporated in atomic basis sets by adding one or more orbitals from a higher level to the valence orbitals of the original set. Polarized basis sets are usually denoted by adding one or two * (star) to their names. For example, basis set 6-31G* implies that d functions have been added to the p functions in set 6-31G. A second star shows that p functions have been added to s functions of H and He atoms.

In addition to polarization functions, *diffuse* functions may be required for anions, molecules with a negative charge. In anions, the wave function of the outermost electron has a long tail. Mathematically, this means that one needs to add a Gaussian function with a small exponent to the basis function. The presence of diffuse functions is often denoted by adding one or two + (plus) signs to the name of the original basis set. For instance, 6-31+G implies that there are one additional s and one additional set of p functions in the original set 6-31G. A second plus, such as in 6-31++G, indicates that a diffuse s functions has been used for the atoms on the first row. As a rule of thumb, the diffuse exponents are about a factor of four smaller than the smallest exponent in the original set. The above mentioned notation is now standard for the Gaussian basis sets of Pople.

Another category of widely used basis sets are those of Dunning [99], called *correlation-consistent polarized n-valence* ζ (cc-pVnZ), where n is the number of ζ exponents. For example, basis set cc-pVTZ implies the use of triple-ζ functions. If diffuse functions are added, the name of the basis set gets prefix *aug-*, such as in aug-cc-pVTZ. The correlation-consistent notion indicates that the contraction coefficients and exponents have been specifically optimized to include electron correlation. The Dunning basis sets have, by default, polarized functions. Unlike Pople's set, they can be expanded without limit in a systematic way to improve the calculation results.

In Dunning's method to develop basis sets, the first primitive function or functions are taken from the Hartree-Fock orbitals, while the rest constitute the most diffuse primitives. The polarization functions are also Gaussian primitives, whose exponents are determined by a uniform scheme called *even-tempered expansions*, given by

$$\zeta_i = \alpha\, \beta^{i-1} \qquad\qquad i = 1, ..., N_f \qquad\qquad (2.106)$$

where N_f is the number of primitives in the set, and α and β are parameters to be optimized for each set. Dunning showed that the correlation effects could be captured efficiently and effectively if this scheme was adopted to find the exponents of Gaussian primitives in an atomic environment.

At present, full electronic treatment of basis sets in molecular environments is practically only possible for light atoms. Heavier atoms in the periodic table possess too many electrons to be handled completely. For example, a neutral platinum atom has 78 electrons, which upon using the minimal basis set STO-3G requires 224 primitive Gaussian functions. Helmann (1935) suggested to replace the nucleus and most of its surrounding electrons by one nuclear point, called *core*. The core would have analytical function that mimics the collective field of the original nucleus and the removed electrons. For this reason, such functions are called *effective core potentials* (ECPs), or *pseudopotentials*.

ECPs are built to match the potential of the core at the radii that are greater than a certain cut-off radius r_c. It is worth noting that using ECPs turns out to be more beneficial than merely simplifying the computation of heavy atoms. The core electrons in heavy atoms usually move at speeds that are comparable to the speed of light. This implies that for these electrons the relativistic effects become significant, and need to be considered. Recall that these effects are not represented in the Schrödinger equation (2.7) because of the non-relativistic Hamiltonian. ECPs offer an alternative path to consider relativistic effects, because they have the potential to exactly represent such effects at a cheap computational cost.

A popular set of ECPs are those of Hay and Wadt [49, 104], also known as Los Alamos National Laboratory (LANL) pseudopotentials. Currently, two flavors of this basis set are in common use: LANL single-ζ and LANL double-ζ. The LANL basis sets do consider the relativistic effects for heavy elements. In particular for transition metals, they define the valence orbitals for the first, second, and third series to be respectively $(3d,4s,4p)$, $(4d,5s,5p)$, and $(5d,6s,6p)$. The core orbitals are then represented by effective potentials as

$$
\begin{aligned}
1^{st} \text{ transition row}: &\qquad ...3s^2\,3p^6 \equiv [\text{Ar}], \\
2^{nd} \text{ transition row}: &\qquad ...3d^{10}\,4s^2\,4p^6 \equiv [\text{Kr}], \\
3^{rd} \text{ transition row}: &\quad ...4d^{10}\,5s^2\,5p^6\,4f^{14} \equiv [\text{Xe}]\,4f^{14}
\end{aligned}
\tag{2.107}
$$

The notation in brackets indicates the configuration of the corresponding noble gas: Ar, Kr or Xe.

Most basis sets are based on analytical expressions both in their development and in their implementations. There exist also purely numerical and/or empirical basis sets having special computational or application features. In the following, we will focus on the comparison of different computational theories when combined with different basis sets.

2.6   Comparison of Performance and Accuracy

Two main aspects of any quantum calculation that determine the computation time and the accuracy of the results are the level of theory and the applied basis sets. The quality of both aspects, and in particular, their suitable combination are crucial factors to the success of the predicted properties. The properties of interest range from structural properties, such as equilibrium/non-equilibrium geometries and vibrational frequencies, to energetics such as total energies, activation energies, and electron affinities. One should not generally expect highly accurate results from a high level of theory if combined with a poor basis set, or vice versa. Obviously, depending on the desired accuracy in the computed quantity, one must choose, among different methods and basis sets. In this section, we present the major issues to be considered when making computational choices.

First, let us consider how calculations scale up with the number of basis sets N.

Table 2.1 shows the formal trends, with HF scaling with N^4. MP stands for Møller-Plesset perturbation methods. QCISD is quadratic configuration interaction including singles and doubles. CCSD is the method of coupled cluster, and T and Q stand for triple and quadruple in the notations used in Table 2.1.

Table 2.1: Scaling behavior of methods versus the number of basis functions N, (Cramer [31])

Method	Scaling
HF	N^4
DFT (plane-wave)	N^3
MPn: $n = 2, ..., 7$	N^{n+3}
CISD, CCSD, QCISD, MP4SDQ	N^6
CISDT, CCSDT	N^8
CISDTQ, CCSDTQ	N^{10}

Note that nowadays the situation is not as restrictive as listed. For instance, in very large systems, the computation of many near-zero integrals can be avoided using a technique called *pre-screening*. Moreover, improving the current computational methods in order to develop so-called *linearly-scaled* methods is still an ongoing topic of research. A few ideas on linear methods have already been implemented in computer programs, and more developments have to be expected.

As the second property, we discuss the accuracy of calculated geometries. In general, the equilibrium geometries are predicted more accurately than the transition geometries. Equilibrium structures calculated at the Hartree-Fock level are in many cases comparable to those obtained by superior density functional or by perturbation methods. Because of correlation errors, HF usually underestimates the bond lengths with respect to the more accurate methods or to the experimental values. Bond angles, on the other hand, show less sensitivity to the level of theory. HF, MP2 and

DFT models all can determine the bond angles on average within say $1°$.

Table 2.2 compares the mean absolute error in bond lengths for different methods using 6-31G** or 6-311G** basis sets. The test set, called G2, comprises of 32 molecules containing only the first row atoms. It is seen that the error introduced by the B3LYP functional is comparably lower than the other methods. B3LYP, being a semi-empirical method, is known as a cost-effective method which gives satisfactory accuracy especially in predicting equilibrium geometries.

Table 2.2: Mean absolute errors in bond lengths for different methods for the G2 set, (Cramer [31])

Level of theory	mean absolute error
MO theoretical methods	
HF	0.022
MP2	0.014
QCISD	0.013
CCSD(T)	0.005
GGA functionals	
BLYP	0.014
BPW91	0.014
Hybrid functionals	
B1LYP	0.005
B1PW91	0.010
B3LYP	0.004

Next, let us focus on the energetics of quantum computations. The energy of a system is definitely the most important quantity in almost all calculations. As a general criterion, computations cannot be considered converged if the energy is not within a satisfactory tolerance. Energies are often more sensitive to both theory and basis set than the structure of molecules. Therefore, because the structure optimization can become too expensive in a high level theory, one may determine a certain structure by a cheaper method, and calculate the corresponding energy using a better method. There exist also interpolation methods, which combine low-cost computations on a system to evaluate highly accurate energetics. The usual expectation is to obtain better estimates as the basis set is improved, but this is not always the case. For example, it is well known that DFT models, despite providing converged results, may not lead to a corresponding accuracy with increasing the size of the basis set.

Table 2.3 gives the mean errors in enthalpies of activation and forward reaction for $HO+CH_4 \rightarrow H_2O + CH_3$, $H+OH \rightarrow O+H_2$, and $H+H_2S \rightarrow H_2+HS$. All the geometries in this set were obtained at QCISD/MG3 level. Thus, the corresponding errors would somewhat vary if the geometries were obtained according to each method. The basis set used in test reactions of Table 2.3 was 6-31+G**. In Table 2.4, the mean absolute

Table 2.3: Mean absolute errors in enthalpies of the test reaction set reported in kcal/mol,(Cramer [31])

Level of theory	error in activation	error in reaction
HF	12.4	149.5
MP2	5.5	24.4
QCISD	3.9	38.6
QCISD(T)	3.1	32.3
KMLYP	2.9	1.0
BLYP	8.3	6.8
B3LYP	5.0	7.2
CBS-Q	0.8	1.3

errors of ionization potentials (IPs) and electron affinities (EAs) for test set G2 are compared.

Table 2.4: Mean absolute errors in IPs and EAs for test set G2 in eV,(Cramer [31])

Level of theory	error in IP	error in EA
MO theoretical methods		
MP2	0.1	0.1
CBS-QB3	0.05	0.05
GGA functionals		
BLYP/6-311+G(3df,2p)	0.7	0.7
BLYP/aug-cc-pVTZ	0.2	0.1
Hybrid functionals		
B3LYP/cc-pVDZ	0.2	
B3LYP/6-311+G(3df,2p)	0.17	0.1
B3LYP/aug-cc-pVTZ	0.2	0.1
X3LYP/6-311+G(3df,2p)	0.15	0.09

2.7 Summary

In this chapter, the fundamentals of quantum chemical calculations were briefly presented. Starting from the Schrödinger wave equation in the time-domain along with the Born-Oppenheimer approximation, different theories were derived to compute the properties of chemical systems. Two main approaches to solve the Schrödinger equation for chemical systems are Molecular Orbital (or Wave Function) methods, and Density Functional methods. There are also hybrid methods that combine both models and may use fitted parameters as well. Basis sets as the other side of quantum computations were also discussed. Finally, we compared the performance of different methods and basis sets in predicting the structure and energetics of test cases. In this work, we have widely used B3LYP, which is a hybrid method with an overall satisfactory performance in terms of computation time and accuracy. In most calculations, the correlation-consistent functions of Dunning, and the relativistic basis sets of Hay and Wadt have been used.

Chapter 3

Electrochemistry in PEM Fuel Cells

In this chapter, we first review the literature on the involved chemistry in PEM fuel cells. Then, a suitable chemical mechanism for the Oxygen Reduction Reaction (ORR) is presented. This chemical mechanism is based on a thorough literature survey [107]. Next we focus on electron transfer reactions in that mechanism and study those reactions from theoretical points of views: both macroscopic and microscopic pictures are discussed. The macroscopic point of view roots in the chemical kinetic concepts, leading to the Butler-Volmer equation. This equation is conventionally used to model the chemistry of fuel cells, which is often mapped to one single global reaction, or a rate determining reaction.

However, the Butler-Volmer equation can also serve as a convenient form to carry the kinetic information of elementary steps, which may be obtained by any approach including experimental measurements or microscopic theories and quantum calculations. The microscopic view is discussed in the framework of the Marcus theory for electron transfer. This theory has been primarily applied to outer sphere reactions, reactions in which the electron donor and acceptor interact through a long distance compared to the size of the molecules. Through decades of application to real systems, Marcus theory is thought to predict the correct trends in almost all electron transfer reactions in various fields ranging from biology to photochemistry and electrochemistry.

Fundamental concepts in Marcus theory are presented here because of its close relationship to the Local Reaction Center theory that will be discussed in the next chapter. The theoretical derivations in sections 3.2 and 3.3 are based on text books devoted to electrochemistry and electron transfer. More comprehensive details can be found in Refs. [19, 26, 71, 93].

3.1 Literature Review

As pointed out in Chapter 1, Proton Exchange Membrane (PEM) fuel cells might play a major role in future energy systems for a wide range of applications. However, the performance of PEM fuel cells needs to be significantly improved to create an economic rival to conventional power generation systems. A major obstacle comes from the slow electrochemical reactions at the cathode electrode causing large potential losses in PEM fuel cells. The standard reversible potential for O_2 reduction reaction is calculated to be 1.23 V, but because of slow kinetics, the working voltage is often below 0.8 V, even on platinum electrodes [4]. Understanding the involved chemical phenomena requires a detailed knowledge about the important reactions at the atomic level. To go beyond a single-step representation of the electrochemistry in a PEM fuel cell, a suitable reaction mechanism has to be formulated which includes both chemical and electrochemical (electron transfer) steps. The two main aspects of this are to establish a reaction mechanism and to estimate kinetic rate data.

The factors affecting the oxygen reduction reaction have been reviewed recently by Markovic et al. [63, 64], Adzic et al. [1] and other groups. The focus of most of the work on the oxygen reduction reaction (ORR) processes occurring on Platinum electrodes has been the experimental characterization and analysis. Experimental data are available for the dependence of ORR kinetics on various environmental parameters such as pressure, temperature, humidity, reactant concentrations, and electrolytes [21, 74, 82, 83, 106, 112, 119]. Empirical models are obtained based on these results for representing the overall system dynamics. A relatively detailed understanding of the fundamental reactions has been obtained from these studies and was further complemented by computational studies of several researchers [7, 8, 11, 46, 57, 58, 69, 75, 76, 91, 100, 117, 118].

Different reaction pathways exist for the overall oxygen reduction reaction [64]. On platinum electrodes, O_2 reduction reaction is considered to proceed either through a direct electroreduction to H_2O, or through a parallel pathway with H_2O_2 as an intermediate species [114]. However, the direct reduction pathway is more prominent at the working voltage regime [62]. The overall reduction reaction is inherently irreversible even on platinum electrodes [64]. The efficiency is further reduced by the simultaneous occurrence of parasitic reactions which compete in establishing a mixed potential [80, 111]. The oxidation of adsorbed impurities that are present in the electrolyte may constitute a possible side reaction, which results in a parasitic anodic current [111]. Others, like the corrosion reaction, are considered to be the source of adsorbed species, which block the active sites for oxygen reduction [14, 80]. In addition to the surface blocking effect, the interactions between the adsorbed species are also shown to affect the reaction kinetics [108].

The oxygen reduction reaction includes a series of single electron transfer reactions intermediated by a few chemical steps which constitute the overall mechanism. Many authors believe that the first electron transfer reaction is the rate determining step for oxygen reduction in acidic media [7, 81], while others attribute the overall rate to a chemical reaction [32, 114]. The idea of a rate determining step (RDS) implies that other reactions are very facile as compared to the RDS, and hence can be considered in a quasi-equilibrium state [32]. Because H_2O is a key species in the overall reaction, the mechanism has also been suggested to be affected by the relative humidity values for the Nafion electrolytes [112]. Nevertheless, the formulation of a reliable reaction mechanism requires investigating all individual steps. The first group of reactions are thermal reactions on the Pt(111) and (100) surfaces. This set includes the dissociation of O_2, O_2H, and H_2O as well as two hydrogen-exchange reactions on catalytic Pt surfaces. The second set includes electron transfer reactions, and the third set contains corrosion steps. We present here a compiled mechanism in Table 3.1 that contains the important elementary steps of oxygen reduction on Platinum electrodes. This table covers thermal, electrochemical, and corrosion reactions, which have been extracted from a literature survey in our group [107].

There exist in the literature comprehensive studies on the thermal reactions using computational tools. Michaelides and Hu [69] studied the dissociation of water, recombination of O and H, and recombination of OH and H on the Pt(111) surface using a plane wave DFT method. They also studied the bonding of OH on a Pt(111) surface in on-top, bridge, and 3-fold sites [68]. Li and Balbuena [58] simulated the dissociation of O_2 on a Pt_5 cluster, and mainly focused on the electrochemical aspect. Wang and Balbuena [109] considered the dissociation of O_2 on a Pt cluster using molecular dynamics simulations, and found that the dissociation proceeds via an asymmetric path.

The other category of reactions consists of electron transfer or electrochemical reactions. In recent years, quantum computations have also been performed to calculate the energetics of electrochemical reactions [11, 35] using different methods and basis sets. The side reactions, on the other hand, are suggested to be responsible for some deviations of measurements from theoretical estimations or usual expectations.

As the focus of this book is the kinetics of elementary electron transfer reactions in Table 3.1, here we discuss only the theories and models that are related to these. The kinetics of thermal reactions are discussed in [52, 107], while the side reactions still have to be investigated in more details through future studies.

Table 3.1: Reaction steps constituting the oxygen reduction reaction at cathode electrode of PEM fuel cells.

Reaction type	Reaction	Reference
Chemical	$O_{2(ads)} \rightleftharpoons O_{(ads)} + O_{(ads)}$	[114], [95], [7]
	$O_2H_{(ads)} \rightleftharpoons O_{(ads)} + OH_{(ads)}$	[33], [32], [111], [34]
	$H_2O_{(ads)} \rightleftharpoons H_{(ads)} + OH_{(ads)}$	[11]
	$OH_{(ads)} + OH_{(ads)} \rightleftharpoons O_{(ads)} + OH_{2(ads)}$	[39], [3], [69]
	$OH_{(ads)} + O_{(ads)} \rightleftharpoons O_{(ads)} + OH_{(ads)}$	[107]
Electrochemical	$O_{2(ads)} + H^+_{(aq)} + e^- \rightleftharpoons O_2H_{(ads)}$	[33], [32], [111], [34]
	$OH_{(ads)} + H^+_{(aq)} + e^- \rightleftharpoons OH_2$	[114], [111], [105], [108]
	$O_{(ads)} + H^+_{(aq)} + e^- \rightleftharpoons OH_{(ads)}$	[114], [111], [33], [32]
Side Reactions	$Pt \rightleftharpoons Pt^{2+}_{(aq)} + 2\,e^-$	[38], [113], [94]
	$Pt^{2+}_{(aq)} + H_2O \rightleftharpoons PtO + 2\,H^+_{(aq)}$	[38], [113], [94]
	$Pt^{2+}_{(aq)} + H_2 \rightleftharpoons Pt + 2\,H^+_{(aq)}$	[38], [113], [94]

3.2   Macroscopic View in Electron Transfer

Any charge migration involves the transfer of at least one electron between two ions in solution or between an ion and an electrified surface. Electrochemistry in general refers to charge transfer processes that occur on electrified surfaces or electrodes. All the electrochemistry of PEM fuel cells, as has been compiled in Table 3.1, consists of charge transfer on electrodes. Because the simultaneous transfer of more than one electron is very unlikely to happen, it is reasonable to consider all electrochemical reactions as single one-electron transfer steps. The term electrochemical and electron-transfer (ET) will thus be used interchangeably in this book.

Real charge transfer between an ion and an electrode can be accompanied by complex changes in the atomic structure of the system, such as breaking/formation of bonds and adsorption/desorption of molecules. In addition, catalysts and previously adsorbed layers are often present, and can play a crucial role. Usually ET reactions are categorized as being *outer-sphere* or *inner-sphere* depending on the physical nature of charge transfer. An electron transfer process is characterized as outer-sphere if

- interatomic bonds do not change appreciably,

- adsorption or desorption do not take place,

- catalysts do not play any significant role.

Otherwise, the reaction is considered as an inner-sphere ET process. In outer-sphere reactions, the electron transfers between the electrode and an ion that is far away from the electrode, and often solvated in the solution. Well-known outer-sphere reactants are typically metal ions surrounded by inert ligands that keep the core metal from directly interacting with the electrode. For example, reaction

$$[Fe(H_2O)_6]^{2+} \rightleftharpoons [Fe(H_2O)_6]^{3+} + e^- \tag{3.1}$$

is known to proceed via an outer-sphere path in the absence of halide catalysts. Being isolated from the electrode, outer-sphere reactions proceed at rates which are almost independent of the electrode, its exact structure and catalytic properties. The theory and kinetics of outer-sphere reactions are well understood and documented in the literature.

In comparison, inner-sphere reactions exhibit more complexity, which is the main reason why concepts, theories, and models describing this category still witness new ideas and developments. Although most electron transfer reactions that are of great importance in technology and science fall into the inner-sphere category, it is important to study outer-sphere reactions as the ideal or extreme cases of inner-sphere processes.

Consider the one-electron reaction

$$\text{Red} \underset{k_{\text{red}}}{\overset{k_{\text{ox}}}{\rightleftharpoons}} \text{Ox} + e^- \tag{3.2}$$

which proceeds with rates k_{red} and k_{ox} in the reduction and oxidation directions, respectively. Denoting the concentration of the reduced (Red) and oxidized (Ox) species at the electrode surface by c^s_{red} and c^s_{ox}, the net anodic current density j is given by

$$j = F \left(k_{\text{ox}} \, c^s_{\text{red}} - k_{\text{red}} \, c^s_{\text{ox}} \right) \tag{3.3}$$

with F being the Faraday constant. As discussed in section 1.5, k_{red} and k_{ox} can be expressed by the following Arrhenius equations within the assumptions of the transition state (TS) theory

$$
\begin{aligned}
k_{\text{red}} &= A \, \exp\left(-\frac{\Delta G^\dagger_{\text{red}}(\phi)}{RT} \right) \\
k_{\text{ox}} &= A \, \exp\left(-\frac{\Delta G^\dagger_{\text{ox}}(\phi)}{RT} \right)
\end{aligned}
\tag{3.4}
$$

In Appendix B, it is shown that the assumption of equal pre-exponential factors is valid, at least, at standard and equilibrium conditions. Note also that ΔE has been replaced here by ΔG for future theoretical derivations to emphasize it is the change in the Gibbs free energy. The activation energies $\Delta G^\dagger_{\text{red}}$ and $\Delta G^\dagger_{\text{ox}}$ are functions of the electrode potential ϕ, while the pre-exponential factor A presumably is not. The potential at which the forward and backward reactions occur at the same rate is called *the equilibrium* potential ϕ_0. At this potential the activation energies for the reduction and oxidation processes are equal

$$\Delta G^\dagger_{\text{red}}(\phi_0) = \Delta G^\dagger_{\text{ox}}(\phi_0) \qquad @\ equilibrium \qquad (3.5)$$

At potentials more positive than ϕ_0, the oxidation step proceeds faster than the reduction step. This can be simply explained and understood: at more positive potentials, electrons on the electrode are more stable and less willing to leave the surface toward solvated ions. The reverse effect is true for potentials higher than ϕ_0. Apart from this expected phenomenological behavior, a valid question is how the activation energies depend on the electrode potential? Is that dependence linear or non-linear? If a linear dependence is assumed, the Butler-Volmer equation is obtained. To see this, let expand the activation energies about ϕ_0 and keep the first order derivatives only

$$\begin{aligned}
\Delta G^\dagger_{\text{ox}}(\phi) &= \Delta G^\dagger_{\text{ox}}(\phi_0) - \alpha\, F\,(\phi - \phi_0) \\
\Delta G^\dagger_{\text{red}}(\phi) &= \Delta G^\dagger_{\text{red}}(\phi_0) + \beta\, F\,(\phi - \phi_0)
\end{aligned} \qquad (3.6)$$

where α and β are two positive constants defined by

$$\begin{aligned}
\alpha &\equiv -\frac{1}{F}\left.\frac{\partial \Delta G^\dagger_{\text{ox}}}{\partial \phi}\right|_{\phi=\phi_0} \\
\beta &\equiv \frac{1}{F}\left.\frac{\partial \Delta G^\dagger_{\text{red}}}{\partial \phi}\right|_{\phi=\phi_0}
\end{aligned} \qquad (3.7)$$

The sign of the first order terms in (3.6) have been chosen such that they predicts correct physical trends in the activation energies as a function of potential.

Using expressions (3.6) and (3.4), the current density in equation (3.3) is now equal to

$$j = F k_0 \left[c_{\text{red}}^s \exp\left(\frac{\alpha F (\phi - \phi_0)}{RT} \right) - c_{\text{ox}}^s \exp\left(-\frac{\beta F (\phi - \phi_0)}{RT} \right) \right] \qquad (3.8)$$

with

$$k_0 = A \exp\left(-\frac{\Delta G_{\text{ox}}^\dagger(\phi_0)}{RT} \right) = A \exp\left(-\frac{\Delta G_{\text{red}}^\dagger(\phi_0)}{RT} \right) \qquad (3.9)$$

Equation (3.8) is the Butler-Volmer (BV) equation in its most general form. It shows the basic exponential dependence of current versus voltage in electrochemical cells.

For outer-sphere reactions or assuming such conditions, a simple relationship exists between α and β which leads to the conventional form of equation (3.8). At any potential ϕ, the Gibbs free energies of activation are related to $G_{\text{ox}}^\dagger$ and $G_{\text{red}}^\dagger$ the Gibbs free energies of the oxidized and reduced states through the following equation

$$\Delta G_{\text{ox}}^\dagger(\phi) - \Delta G_{\text{red}}^\dagger(\phi) = G_{\text{red}} - G_{\text{ox}} \qquad (3.10)$$

If the reactants are far enough from the electrode and blocked from it by an inert electrolyte, their electrostatic potential remains almost the same when the potential changes. Under these conditions, which in general are satisfied in outer-sphere reactions, the right hand side of equation (3.10) reflects only the change in Gibbs free energy of electrons in the electrode. If the electrode is now at potential ϕ, this change is nothing more than $-F(\phi - \phi_0)$. Thus one can rewrite equation (3.10) as

$$\Delta G_{\text{ox}}^\dagger(\phi) - \Delta G_{\text{red}}^\dagger(\phi) = -F (\phi - \phi_0) \qquad (3.11)$$

Differentiation of (3.11) with respect to ϕ and using definitions (3.7) yields

$$\alpha + \beta = 1 \qquad (3.12)$$

Being positive numbers, α and β have to lie between zero and one according to equation (3.12). They show how much of the change in the ground state energies of reactants and products is transfered to the energy of the transition state. If they are equal, reduction and oxidation directions take the same share. Based on this interpretation, α and β are called the *symmetry coefficients*.

Using equation (3.12), the BV equation (3.8) is simplified to the form

$$j = F\, k_0 \left[c_{\text{red}}^s \exp\left(\frac{\alpha\, F\, (\phi - \phi_0)}{RT} \right) - c_{\text{ox}}^s \exp\left(-\frac{(1 - \alpha)\, F\, (\phi - \phi_0)}{RT} \right) \right] \qquad (3.13)$$

Now consider the potential at which the net current is zero: equation (3.13) gives

$$\phi = \phi_0 - \frac{RT}{F} \ln \frac{c_{\text{ox}}^s}{c_{\text{red}}^s} \qquad (3.14)$$

If conditions are *standard*, or more generally if $c_{\text{red}}^s = c_{\text{ox}}^s$, the standard equilibrium potential is denoted by ϕ_{00}, which is the value reported in the electrochemical tables for half-cell reactions. Now ϕ_0 at non-standard conditions is given by the Nernst equation as a function of ϕ_{00}

$$\phi_0 = \phi_{00} + \frac{RT}{F} \ln \frac{c_{\text{ox}}^s}{c_{\text{red}}^s} \qquad (3.15)$$

The physical significance of the above relations is explained below.

From equation (3.13), it is seen that at ϕ_0 the net current is zero only if the surface concentrations are equal. Otherwise, the reaction proceeds in such a way to cancel the unbalanced concentrations and arrive at a new equilibrium state. In this case, the reversible or equilibrium potential ϕ_0 has to shift from its standard value ϕ_{00} to a different value to compensate for the change in concentrations according to the Nernst equation (3.15). Therefore compared to ϕ_{00}, ϕ_0 is actually a function of the surface concentrations, which may vary with any experimental setup and any measurement. It is thus desirable to express the BV form (3.13) in terms of ϕ_{00}.

Substituting equation (3.15) in (3.13) yields

$$j = j_0 \left[\exp\left(\frac{\alpha F \eta}{RT} \right) - \exp\left(-\frac{(1-\alpha) F \eta}{RT} \right) \right] \qquad (3.16)$$

with the *overpotential* η defined as

$$\eta \equiv \phi - \phi_{00} \qquad (3.17)$$

and the *exchange current density* j_0 given by

$$j_0 = F k_0 (c_{\mathrm{red}}^s)^{(1-\alpha)} (c_{\mathrm{ox}}^s)^{\alpha} \qquad (3.18)$$

j_0 represents the collective effect on the current density due to surface concentrations and also that part of the kinetics stemming from the vibrations of atoms. The energetic effect in kinetics is represented by η, which exponentially affects the current.

Figure 3.1 illustrates the dependence of current density as a function of η for different values of α. If α is very close to zero, or at high overpotentials, the second term is negligible with respect to the first term in equation (3.16). The net current will thus be almost only anodic. The reverse is true if α is very close to one or at low overpotentials. These two extreme situations, in which the current-voltage relationship has a simple exponential form, are referred to as *Tafel* regions. The logarithm of current versus potential or overpotential in Tafel regions is a straight line. In other cases, both terms in equation (3.16) are comparable in magnitude and contribute to the net current, thus both have to be considered.

For low overpotentials, where $|\eta| \ll RT/F$, both exponential terms in equation (3.16) can be expanded around $\eta = 0$. The expression that results for the current density j is then independent of the value of α, and is equal to

$$j = j_0 \frac{F}{RT} \eta \qquad (3.19)$$

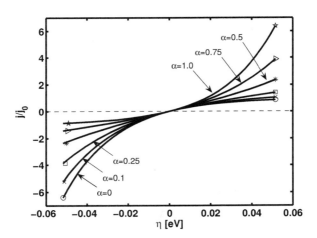

Figure 3.1: Normalized current density versus overpotential at room temperature according to the Butler-Volmer equation (3.16)

Therefore, measurements at small overpotentials cannot be used to determine an accurate value for α. In general, α is assumed to be very close or equal to $1/2$, which gives satisfactory results within the approximations made to derive the Butler-Volmer equation. Due to the use of only first order expansions in (3.6), one should not expect very accurate results from the BV equation at high potentials. Rather it might be necessary to include higher order terms as well to capture the nonlinear effect of the electrode potential on the activation energies. The second order theories of electron transfer are based on the Marcus theory that is discussed in the next section.

3.3 Microscopic View: Marcus Theory

Marcus theory is a semi-classical model that is based on atomic description of reactants and products. The classical aspect enters the theory through ignoring quantum tunneling and other assumptions that are discussed below. This theory was initially developed for outer-sphere processes [60], but later was extended to model inner-sphere reactions too [29]. Despite its simple formulation, the Marcus model has been very successful in estimating rates of electron-transfer, and in particular in elucidating the role of the environment.

The first assumption in Marcus theory is that electron transfer happens via a radiationless process, *i.e.* no energy is exchanged during the electron transfer (ET) with the far away medium. From this point of view, the ET process is assumed to occur via an *isoenergetic* path. The second assumption is based on the Frank-Condon principle: the configuration of the nuclei is the same right before and right after electron transfer. The physical rationale behind this principle lies in the fact that the relaxation time of nuclei motion is much longer than that of electrons. In other words, the nuclei do not find enough time to change their position during the transfer of electron, and the actual reorientation of the nuclei occurs after the electron transfer. The Frank-Condon principle allows one to assume the atomic structure of reactants and products to be essentially the same.

As discussed in Chapter 2, the atomic structure determines the internal energy E (or G) of the system. If the system is initially at the ground state, its potential energy is minimum, implying that all the first derivatives of G with respect to nuclei positions vanish. If now the structure undergoes vibrations and differs from the ground state, G should increase with respect to its minimum G^0. Similar to classical mechanics to model vibrations around a local minimum, G can be expanded around the equilibrium configuration of the structure

$$G_a(\mathbf{q}) = G_a^0 + \sum_i \frac{1}{2} m_i \omega_i^2 (q_i - q_{a,i}^0)^2 \qquad a = \text{ox}, \text{red} \qquad (3.20)$$

where $\mathbf{q} = \{q_i\}$ are the normal coordinates related to the atomic structure, and their equilibrium values are given by $\mathbf{q}_a^0 = \{q_{a,i}^0\}$. m_i and ω_i represent the effective mass and vibrational frequency of normal coordinate q_i, respectively. To simplify the mathematics, in equation (3.20) it has been assumed that the frequencies of the oxidized and reduced centers are equal.

According to equation (3.20), the hypersurfaces G_{ox} and G_{red} are two paraboloids in terms of the normal coordinates. Figure 3.2 demonstrates the potential energy surfaces of the redox couple as a function of reaction coordinate. Assume the system is initially in the oxidized ground state having normal coordinates $\mathbf{q}_{\text{ox}}^0$. Then the nuclei can vibrate because of their thermal energy and due to other possible external excitations. If the system can be excited enough, it can reach a certain configuration $\mathbf{q}^\#$, where the two energy surfaces intersect, that is

$$G_{\text{ox}}(\mathbf{q}^\#) = G_{\text{red}}(\mathbf{q}^\#) \qquad (3.21)$$

If an electron is removed from the system to the electrode, it has to happen when the system lies on the intersection path $\mathbf{q}^\#$. At the same time, the reaction occurs preferably at a point with the lowest energy $G_{\text{ox}}^\#$. Thus finding $\mathbf{q}^\#$ is a minimization problem subject to constraint (3.21). This problem can be formulated and solved

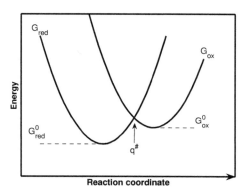

Figure 3.2: Potential energy surfaces of redox centers according to the Marcus theory.

using the Lagrange multiplier method. Consider the Lagrange function

$$\mathcal{L}(\mathbf{q}, \mu) = G_{ox}(\mathbf{q}) + \mu \left[G_{red}(\mathbf{q}) - G_{ox}(\mathbf{q}) \right] \tag{3.22}$$

with μ being the Lagrange multiplier, which enforces condition (3.21).

Using (3.20), the solution is given by

$$q_i^{\#} = q_{ox,i}^0 + \mu^{\#} \left(q_{red,i}^0 - q_{ox,i}^0 \right) \tag{3.23}$$

where $\mu^{\#}$ is equal to

$$\mu^{\#} = \frac{\lambda + G_{red}^0 - G_{ox}^0}{2\lambda} \tag{3.24}$$

λ is called the *reorganization energy*, and is defined as

$$\lambda \equiv \frac{1}{2} \sum_i m_i \omega_i^2 (q_{ox,i}^0 - q_{red,i}^0)^2 \tag{3.25}$$

λ is the energy required to reorient the reactants from their ground state to adopt the ground state configuration of the products without the transfer of the electron, or vice versa.

The activation energies for the reduction and oxidation steps are obtained by subtracting the ground state energies from those at the transition point

$$
\begin{aligned}
E_{\text{red}}^{\text{act}} = \Delta G_{\text{red}}^{\dagger} = \frac{(\lambda + G_{\text{red}}^0 - G_{\text{ox}}^0)^2}{4\lambda} \\
E_{\text{ox}}^{\text{act}} = \Delta G_{\text{ox}}^{\dagger} = \frac{(\lambda + G_{\text{ox}}^0 - G_{\text{red}}^0)^2}{4\lambda}
\end{aligned}
\tag{3.26}
$$

If an ET reaction happens between the reactants and an electrified surface, the energy G of the system depends on the electrode potential ϕ, or equivalently on overpotential η. In that case, the above discussions and Fig. 3.2 are to be considered valid for a certain potential, and they may change for other potential values.

In outer-sphere reactions, the reaction energy ΔG equals the energy that is transfered by the electron, and is related to η according to equations (3.10) and (3.11)

$$
G_{\text{red}}^0 - G_{\text{ox}}^0 = -F\eta
\tag{3.27}
$$

The rate constants in equation (3.4) can be given as a function of the overpotential as

$$
\begin{aligned}
k_{\text{ox}} = A \exp\left(-\frac{(\lambda - F\eta)^2}{4\lambda RT}\right) \\
k_{\text{red}} = A \exp\left(-\frac{(\lambda + F\eta)^2}{4\lambda RT}\right)
\end{aligned}
\tag{3.28}
$$

Equations (3.28) show that the activation energies of electron transfer in Marcus theory are quadratic functions of the overpotential, in contrast to the linear dependence in the Butler-Volmer approximation.

At low overpotentials where $|\eta| << \lambda/F$, one can linearize the activation energies in equation (3.26), and obtain a linear dependence as in Butler-Volmer approximation.

$$\Delta G^{\dagger}_{\mathrm{red}} = \frac{\lambda + 2F\eta}{4}$$
$$\Delta G^{\dagger}_{\mathrm{ox}} = \frac{\lambda - 2F\eta}{4} \tag{3.29}$$

Comparing equation (3.29) with (3.6), one can determine the transfer coefficient α to be 1/2, and the standard activation energy to be $\lambda/4$. If the vibrational frequencies $\{\omega_i\}$ are different for the oxidized and reduced centers, α will be somewhat different from 1/2. Therefore, Marcus theory not only recovers the Butler-Volmer equation at low overpotentials, but also provides an estimation for activation energies at standard potential ϕ_{00}.

An interesting prediction made by the Marcus model is the presence of the so-called *inverted region* at high overpotentials where $|\eta| >> \lambda/F$. Consider an oxidation ET process that is occurring under a very positive overpotential. For such a case, activation energy according to the BV formulation (3.29) should monotonically decrease, and the oxidation rate should increase. In contrast, Marcus theory suggests that if η increases beyond λ/F, the activation energy increases again and the oxidation rate decreases. The reason for this difference is the quadratic dependence of activation energy on η in the Marcus model, equation (3.28), versus the linear dependence in the Butler-Volmer model, equation (3.29). The inverted region has been experimentally observed for many thermal electron transfer processes such as neutralization of ion pairs [97].

Despite ongoing controversies about the inverted region, R.A. Marcus received the Nobel Prize in Chemistry in 1992 for his theory. Other than predicting trends, the Marcus theory is the principal framework for computational studies of electron transfer in various fields.

3.4   Summary

In this chapter, a chemical mechanism was presented for the reactions that occur on the cathode electrodes of PEM fuel cells. A comprehensive literature study suggests that this mechanism must be able to grasp the fundamental phenomena through its many elementary steps. Three reactions in the proposed mechanism are electron transfer processes that occur on the surfaces of catalyst metals such as Platinum.

The theory of electron transfer was discussed from both macroscopic and microscopic points of view. The goal of this research is to provide the most accurate rate data that are available to simulate the electrochemistry of PEM fuel cells. Achieving this goal requires implementing a microscopic theory, here the Marcus theory, of electron transfer using computational tools. The fundamentals for this were discussed in Chapter 2. Therefore, after the theoretical review on electron transfer in this chapter, we focus in the next chapter on a computational model that allows one to determine the transition state of ET processes using quantum chemistry software. Transition state calculations can directly provide activation energies for electrochemical reactions.

Chapter 4

Transition States of ET Reactions

In this chapter, a computational method and an efficient mathematical framework are presented which enable one to determine the potential-dependent transition states of electron transfer (ET) reactions by quantum calculations. Locating transition states is made possible through the theory of Local Reaction Center (LRC) for ET processes. The original implementation of the LRC theory relies on performing several quantum mechanical calculations to locate the transition states. As the size of the system increases, this procedure soon causes the computational cost to limit the LRC theory from being applied to large systems, those having many degrees of freedom.

We develop here an efficient mathematical algorithm that maps the search in an N-dimension space to a 1-dimensional space, and thereby removes the limited application of the LRC theory. It is shown that the new formulation regenerates previously published results obtained by the pattern search approach and the constrained variation method. Our solution algorithm replaces the constrained optimization problem defined in a multidimensional space by a single equation in terms of only one variable that is solved for in each iteration. This method leads to fast convergence, reliability, and robustness of the located transition states for more complex systems with a larger number of degrees of freedom, especially for smooth energy surfaces. The material of this chapter is based completely on one of our publications (Reference [16]).

4.1   Local Reaction Center Theory

Historically, ET processes are discussed in terms of donor and acceptor centers. In this picture, it is assumed that an electron which is initially localized in the donor region is transferred at the transition state to the distinct region of the acceptor [73]. Anderson *et al.* [4, 6], on the other hand, introduce a reaction center that, instead of the conventional donor-acceptor centers in ET theories, communicates with the metal electrode. In this theory, the ET state is defined as the state where the ionization potential (IP), for an oxidation reaction, or the electron affinity (EA), for a reduction reaction, of the reaction center equals the thermodynamic work function of the electrode. The work function of the electrode is in turn a function of the electrode potential E_e.

Denoting EA or IP as ψ, this condition may be written for both oxidation and reduction reactions as

$$\psi \;=\; eE_e \tag{4.1}$$

Equation (4.1) expresses the ET condition in the LRC theory. On the scale of the standard hydrogen electrode (SHE), the electrode potential E_e is given as [7, 78]

$$E_e/\mathrm{V} \;=\; U/\mathrm{V} + 4.6 \tag{4.2}$$

where 4.6 V is the average value of the thermodynamic work function of the standard hydrogen electrode (SHE) based on experimental estimates, and U is the electrode potential with respect to the SHE. The EA and IP of the reaction center are by definition the change in the energy when the system jumps from its initial state to its final state, i.e. the states just before and right after the charge transfer. Because the potential energy surface (PES) of the molecule is a function of multidimensional structure coordinates $\mathbf{x}$ for any given electrode potential, there may exist multiple configuration points on the PES that satisfy the ET condition (equation 4.1). As a result, the transition state of the oxidation/reduction reactions $\mathbf{x}^*$ is identified as the point in the set of solutions satisfying the ET condition, where the activation energy attains its minimum. The activation energy at the transition state φ is the increase in the energy of the system E at the transition state with respect to the energy at the ground state

$$\varphi = E_{\text{transition state}} - E_{\text{ground state}} \tag{4.3}$$

In general, the dependence of the potential energy on $\mathbf{x}$ is not a known function, and quantum computations usually determine only sampled points on the energy surfaces. One approach that has been used to find the transition states with the above theory is the pattern search technique that compares several candidate points on the PES, and picks the one with the lowest activation energy corresponding to a certain electrode potential [10]. Obviously, the sampling process of the PES in the pattern search requires many quantum simulations, and therefore, is a computationally expensive solution approach.

Kostadinov and Anderson [55] used the constrained variation (CV) method, explained below, in order to reduce the number of required quantum simulations. Their approach, based on the Lagrange multipliers method, adopts a more efficient search algorithm to accelerate the determination of transition states. In this algorithm, a new variable μ is introduced to include the ET condition equation (4.1) in the Lagrange function $\mathcal{L}$ defined by

$$\mathcal{L}(\mathbf{x}, \mu) = \varphi(\mathbf{x}) - \mu[\psi(\mathbf{x}) - eE_e]. \tag{4.4}$$

The minimization of $\mathcal{L}$ is achieved if the partial derivatives of L with respect to all independent variables $\mathbf{x}$ and μ are equal to zero. This leads to the following two conditions

$$\nabla \varphi = \mu \nabla \psi, \tag{4.5}$$

$$\psi = eE_e . \tag{4.6}$$

Equation (4.5) is a vector equality, and demands that the gradients of φ and ψ at the transition states be co-linear. The scalar equation (4.6) is the ET condition and a restatement of equation (4.1).

The computational process of the Lagrangian formulation by the CV method is carried out in two steps. In the first step, a point on the PES is found that satisfies the ET condition, i.e. equation (4.6). Next, a particular line search algorithm is followed to maintain the direction of the change in $\mathbf{x}$, the structure variables, perpendicular to the gradient of ψ until the constrained minimum of φ is found by fulfilling equation (4.5) (see Figure 4.1). It is assumed that the ET condition will remain the least perturbed while the algorithm simultaneously attempts to minimize the activation energy. Although the gradients of φ and ψ are calculated by finite differences using a small step size of 0.0004 Å, a shift in potential is observed as the algorithm converges to the corresponding transition states [55]. Therefore, at each step, the final transition state has an IP (or EA) that is slightly different from the target work function of the electrode (Figure 4.1). Despite the substantial improvement of the CV method over the pattern search algorithm, still a large number of quantum simulations are required to compute the gradients and to find the optimum points in the multidimensional space of the configuration variables. As a result, the whole calculation time can increase tremendously if more structure variables of the nuclei configuration participate in the search algorithm.

4.2   Convergent Iterative Constrained Variation Method

In the Born-Oppenheimer approximation, the potential energy surface of a quantum system is a function of the nuclei coordinates or, from a geometrical point of view, a function of the structure variables $\mathbf{x}$. Hence, φ and ψ of the reaction center are also, by definition, functions of $\mathbf{x}$. Consequently, the second order expansions for φ and ψ within the Born-Oppenheimer approximation can be represented as truncated multivariable Taylor series about any local point on the PES

$$dx = x - x_0 \tag{4.7}$$

$$\psi(x) = \psi_0 + B_\psi^T dx + \frac{1}{2} dx^T A_\psi dx \tag{4.8}$$

$$\varphi(x) = \varphi_0 + B_\varphi^T dx + \frac{1}{2} dx^T A_\varphi dx \tag{4.9}$$

where A_ψ, A_φ and B_ψ, B_φ stand for the second and first derivatives of ψ and φ with respect to x evaluated at x_0

$$\begin{aligned} A_\psi &= \left.\frac{\partial^2 \psi}{\partial x \partial x}\right|_{x_0}, & A_\varphi &= \left.\frac{\partial^2 \varphi}{\partial x \partial x}\right|_{x_0} \\ B_\psi &= \left.\frac{\partial \psi}{\partial x}\right|_{x_0}, & B_\varphi &= \left.\frac{\partial \varphi}{\partial x}\right|_{x_0} \end{aligned} \tag{4.10}$$

The first and second order derivatives of the PES are available from the output of single quantum computations without performing additional calculations necessary in finite difference estimations. The gradients of ψ and φ now can be computed by taking the derivative of the above two equations with respect to x

$$\nabla\psi = B_\psi + A_\psi dx, \tag{4.11}$$

$$\nabla\varphi = B_\varphi + A_\varphi dx, \tag{4.12}$$

where $\nabla\psi$ and $\nabla\varphi$ are the first derivatives of ψ and φ with respect to x at any point x around x_0. The desired change dx in structure variables is then expressed in terms of the undetermined parameter μ using Eqs. (4.5), (4.11) and (4.12)

$$dx = (A_\varphi - \mu A_\psi)^{-1}(\mu B_\psi - B_\varphi). \tag{4.13}$$

Equation (4.13) can now be used to eliminate dx in Eqs. (4.11) and (4.12). Substituting the updated expansion of $\nabla\psi$ and $\nabla\varphi$ in equation (4.6) then leads to an equation that is a function of μ only

$$
\psi_0 + \mathbf{B}_\psi (\mathbf{A}_\varphi - \mu \mathbf{A}_\psi)^{-1} (\mu \mathbf{B}_\psi - \mathbf{B}_\varphi) +
$$
$$
+ \frac{1}{2} (\mu \mathbf{B}_\psi - \mathbf{B}_\varphi)^T (\mathbf{A}_\varphi - \mu \mathbf{A}_\psi)^{-T} \mathbf{A}_\psi (\mathbf{A}_\varphi - \mu \mathbf{A}_\psi)^{-1} (\mu \mathbf{B}_\psi - \mathbf{B}_\varphi) = eE_e. \qquad (4.14)
$$

Any root-finding algorithm such as the Newton-Raphson method can be used to solve this scalar equation for μ, which in turn directly leads to the estimation of $d\mathbf{x}$ from equation (4.13). Thus, the CICV method replaces the two-step multi-variable approach of the CV method with a single one-variable step. The increase in the computational cost of this step with increasing the degrees of freedom of a given system is negligible and reflected only in the matrix operations of equation (4.14) and not in the quantum calculations. Thereby, the solution algorithm in the CICV method is, to a great extent, independent of the dimension of $\mathbf{x}$. This will be shown below.

The iteration loop begins with assigning a new value to the electrode potential starting from an initial nuclear configuration $\mathbf{x}_0$. After solving equation (4.14) for μ and equation (4.13) for $d\mathbf{x}$, a candidate set of structural coordinates $\mathbf{x}_1$ is estimated by

$$
\mathbf{x}_1 = \mathbf{x}_0 + d\mathbf{x}. \qquad (4.15)
$$

Next, the EA (or IP) of the system is computed by performing two quantum simulations at the new geometry, which also provide the derivatives required for the Taylor expansions of ψ and φ in Eqs. (4.8) and (4.9). The algorithm loop repeats until the cosine squared of the angle between the gradient vectors $\nabla\varphi$ and $\nabla\psi$, equation (4.16), reaches a desired value close enough to unity.

$$
\cos^2 \alpha = \frac{(\nabla\varphi . \nabla\psi)^2}{\nabla\varphi^2 \nabla\psi^2} \qquad (4.16)
$$

This co-linearity between $\nabla\varphi$ and $\nabla\psi$ assures the fulfillment of equation (4.5) while at the same time, the solution of equation (4.14) in each iteration implies that the EA (or IP) of the reaction center is very close to the work function of the electrode potential.

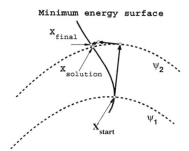

Figure 4.1: Graphical demonstration of the CV algorithm: the CV method locates x_{final}, which corresponds to a slightly different IP (or EA) from the target work function of the electrode ψ_2

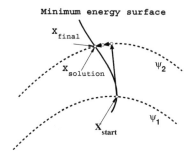

Figure 4.2: Graphical demonstration of the CICV algorithm: the CICV method solves the Lagrange equations such that at each iteration the IP (or EA) of the molecule converges to ψ_2

Because this algorithm solves the Lagrange equations (4.6) and (4.5) at each iteration, it provides more control over the electrode potential at transition states (Figure 4.2). The shift in the electrode potential at transition states observed in the CV method is practically removed, since the tolerance in the target electrode potential can be set to any desired value. The number of iterations needed to locate the transition states within the accuracy of the quantum computations and of the proposed model for ET depends on the smoothness of the PES, *i.e.* on the precision of the second order representation of the Hamiltonian in terms of the structure variables. Hence, for smooth energy surfaces, the CICV method converges almost independently of the dimension of the search space $\mathbf{x}$. This allows a larger number of degrees of freedom in the molecule structure to be included in the constrained optimization of the system at a low computational cost, which is a valuable feature for studying larger systems.

Case Study: Oxidation of PtOH$_2$

For comparison and demonstration purposes, the oxidation of PtOH$_2$ previously studied by Kastadinov and Anderson [55] was investigated and the activation energy versus the electrode potential was obtained using the CICV method presented in the previous section. Oxidation reactions describe the deprotonation of the adsorbed molecule according to

$$\text{PtOH}_2 \rightarrow \text{PtOH} + \text{H}^+ + e^-(U) \tag{4.17}$$

The details of quantum computations, repeated here for convenience, are the same as those used by Kostadinov and Anderson [55]. The hydronium ion was modeled by a hydrogen atom attached to three water molecules, $\text{H}^+\text{-OH}_2(\text{OH}_2)_2$, as shown in Figure 4.3. Five intermolecular distances were varied in the simulation whereas the other distances and angles were fixed at the optimized structure. The contribution of anions and cations was accounted for in the Hamiltonian of the system by adding a point charge of $-\frac{1}{2}e$ placed 10 Å away from the first oxygen atom of the hydronium ion along the H$^+$-O line (see Figure 5.1). This point charge corresponded to the Madelung sum for a rock salt structure of ions in a 0.1M solution of monoprotic acid. The quantum simulations were performed using the MP2 level of theory implemented in GAUSSIAN 03 [42] along with the 6-31G** basis set for O and H, and an effective core potential and double zeta valence orbital basis (LANL2DZ) set for Pt. Reaction center for oxidation of PtOH$_2$.(H$_2$O)$_3$ along with five interatomic distances used to

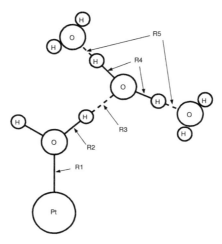

Figure 4.3: Reaction center for oxidation of $PtOH_2.(H_2O)_3$ along with five interatomic distances used to optimize the transition states. R4 and R5 distances were changed symmetrically for both of the two hydrogen-bonded water molecules

optimize the transition states. R4 and R5 distances were changed symmetrically for both of the two hydrogen-bonded water molecules.

Before the actual CICV calculation, the five distances in the reaction center were optimized while the other structure distances and angles were fixed at their corresponding ground state values. The optimized structure is equivalent to the point of zero activation energy illustrated in Figure 4.4, and it serves as the starting point for the CICV computations. The calculated electrode potential at zero activation energy is 0.856 V, which is very close to the value of 0.867 V obtained in Reference [55]. Figure 4.4 demonstrates the potential-dependence of the activation energy for the oxidation of the $PtOH_2$ precursor hydrogen-bonded to three water molecules using the CICV method. It is observed that the CICV method can reproduce the activation energy curves in very good agreement with those obtained by the CV method.

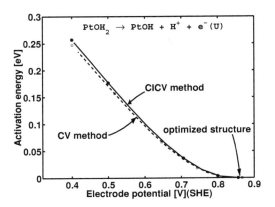

Figure 4.4: Activation energy at different potentials for oxidation of $PtOH_2.(H_2O)_3$ from the CICV method compared with the CV method

The slight disagreement might be due to small differences in the optimized structure parameters.

4.3 Principle of Microscopic Reversibility

To further examine the accuracy of the computed transition states by the CICV method, oxidation/reduction steps of Pt_2O_2H were studied as an additional test case

$$Pt_2 - O_2 + H^+ + e^-(U) \rightleftarrows Pt_2 - O_2H. \tag{4.18}$$

The structure of Pt_2O_2, hydrogen-bonded to a hydronium ion (Figure 4.5), constitutes the precursor for reaction (4.18). Here, we will show that the results predicted by the CICV method satisfy the principle of microscopic reversibility, which requires that the transition configurations be the same at any electrode potential for the forward and backward reactions [8]. Microscopic reversibility offers a way to derive the oxidation curve from the reduction data and vice versa. The agreement between the calculated and the derived curves indicates the consistency of the transition states identified by the solution algorithm. The solution algorithm locates the transition

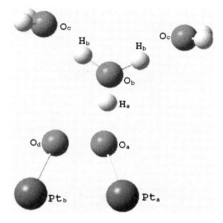

Figure 4.5: The reaction center for Pt_2O_2H oxidation/reduction reactions: Pt_2O_2H attached to three water molecules to account for the effect of the solution

states of the oxidation/reduction steps for the molecule shown in Figure 4.5. The activation energies that are computed from these transition states are called calculated activation energies. On the other hand, the EA of the reduced center and the IP of the oxidized center in equation (4.18), which equal to the electrode potential,

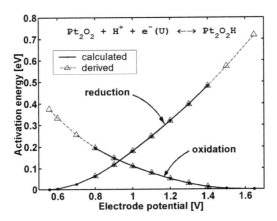

Figure 4.6: Calculated and derived activation energy curves for Pt_2O_2H reactions after the first series of computations using the CICV method to determine the transition states

are determined from the PES of the same precursors. These PES surfaces that are determined independently from both the forward and backward steps, should match in principle. Therefore, at any given electrode potential, the transition state on these PES surfaces should correspond to the same molecular structure. Having computed the transition states from one step, one can obtain the activation energies for the other step using equation (4.3). Here, these latter activation energies are called derived activation energies.

Anderson *et al.* [8] report difficulties in obtaining Pt_2O_2H activation energy curves in the first series of usual calculations by the CV method. Namely, the activation energies obtained by computing the reduction reaction do not match those computed from the data obtained in computing the oxidation reaction, thereby violating the principle of microscopic reversibility. This problem has not been observed for simpler systems, such as hydrogen reactions on Pt, where the transition configurations were found to be compatible for the forward and backward reactions. It was suggested to determine the true transition states of Pt_2O_2H reactions by starting the CV calculations from a structure corresponding to the average of the structures found from the computations of the oxidation and the reduction reactions [8]. Using this procedure, the activation energies obtained after three such iterations were in agreement with those determined from the data of the reverse reaction.

In contrast to the CV method, the activation energy curves from the CICV method are directly obtained in the first calculation, and are in a remarkable agreement with the curves derived from the reverse reactions. This is shown in Figure 4.6. The requirement of microscopic reversibility is hence satisfied, which confirms the reliability of the transition states computed using the CICV method. Ten degrees of freedom were optimized in the constrained variation calculations (Table 4.1). Varying this large number of optimization parameters, which was made possible by the new method, is the reason for observing such a good agreement. The intersection point of the oxidation and reduction activation energy curves in Figure 4.6 corresponds to an electrode potential of 0.93 V, which is only 0.01 V different from the value computed by Anderson *et al.* [8]. The CICV curves for this case also compare very well with the final activation energies of the CV curves, which are not shown here.

4.4 Convergence of the CICV Method

In the solution algorithm of the CICV method, the electrode potential appears as a parameter. Hence, the activation energy has to be computed for each specified value of the electrode potential. The simulations starts typically with the structure of the transition state determined for the previous potential. Within the computation of the activation energy of each specified potential, several iterations may be needed to meet the convergence criteria. A simulation is considered converged if both of the convergence criteria, corresponding to the conditions given by Eqs. (4.5) and (4.6) are met. The first condition is that $1 - \cos^2 \alpha$ has to be smaller than a prescribed tolerance. The second condition demands that $|\psi/e - E_e|$ has to be smaller than a second prescribed tolerance.

Here, the number of iterations required to determine the activation energy for a given potential is used to assess the convergence of the CICV method for different degrees of freedom in the structure variables. The activation energy curve for the oxidation of Pt_2O_2H, given in equation (4.18), was obtained allowing for two, six, and ten degrees of freedom. These degrees of freedom are defined in Table 4.1. For each

Table 4.1: Degrees of freedom of the precursor in the Pt_2O_2H oxidation reaction (4.18)

DOF number	DOF type	Atoms[†]
1	bond-length	H_a-O_a
2	bond-length	O_b-H_a
3	bond-length	O_a-O_d
4	bond-angle	O_a-O_d-Pt_b
5	bond-angle	H_a-O_a-O_d
6	dihedral-angle	H_a-O_a-O_d-Pt_b
7	bond-length	O_d-Pt_b
8	bond-angle	O_d-Pt_b-Pt_a
9	bond-length	H_b-O_b
10	bond-length	O_c-O_b

[†] The atomic labels of the molecule are depicted in Figure 4.5

case, eight different potentials were computed with 0.0005 as the tolerance for the $\cos^2 \alpha$-condition, and 0.01 V for the electrode potential condition. Figure (4.7) shows that more iterations are needed for 6-DOF and 10-DOF cases compared with the 2-DOF case. However, the 10-DOF case needs essentially the same number of iterations to achieve convergence as the 6-DOF system. This result is expected, because the solution algorithm is to a large extent independent of the dimension of the structure variable vector $\mathbf{x}$. In each iteration, only the non-linear scalar equation (4.14) is solved for μ. Consequently, the convergence is mainly affected by the accuracy of the Taylor expansions in Eqs. (4.8) and (4.9) to represent the PES of the system. This depends less on the number of structure variables than the point on the PES used

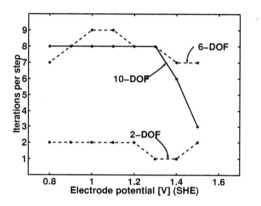

Figure 4.7: Number of Iterations per step required for the oxidation reaction of Pt_2O_2H

as the initial guess. In principle, it could be expected that the expansion is more accurate for a smaller number of degrees of freedom, but the opposite could be the case. As an example, this can be observed for the high potential, 10-DOF simulations, which require fewer iterations than the corresponding 6-DOF simulations.

4.5 Summary

In this chapter, the convergent iterative constrained variation (CICV) method was proposed to compute transition states and potential-dependent activation energies of electron transfer reactions. It was shown that the method offers faster convergence and better control over the desired electrode potential compared to the constrained variation (CV) approach by Kostadinov and Anderson [55] and the pattern search methods. The CICV approach is a constrained optimization method using the method of Lagrange multipliers, which leads to two $(N+1)$ conditions, where N is the number of structure variables. These conditions are combined using Taylor series expansions to a single nonlinear equation for the Lagrange multiplier. The method exploits the first and second order gradients of the potential energy surface, typically provided by quantum simulation packages, such as GAUSSIAN, at no extra computational cost. The calculated activation energy curve for the oxidation of $PtOH_2$ is in very good agreement with the values obtained by the CV method. The CICV method was found to be convergent in the sense of satisfying the microscopic reversibility requirement, even for complex systems. This is demonstrated for oxidation/reduction reactions of Pt_2O_2H. The convergence of the method is shown to be almost independent of the degrees of freedom in the structure variables. Instead, the convergence depends on the accuracy of the second order Taylor series expansion of the PES. The CICV method is used in Chapter 6 to calculate the activation energies of electron transfer reactions. The CICV algorithm has been implemented in a PERL script, VOLMER05 [17], that uses Gaussian program for quantum calculations. VOLMER05 is freely available from the author.

Chapter 5

Computational Investigation of LRC Theory

Having explained the LRC theory and its computational implementation in chapter 4, the questions of the accuracy of the presented procedure used to locate the transition states of electron transfer reactions can be asked. Several issues might introduce uncertainties, most notably the effect of the Madelung charge on the computations, the significance and accuracy of ionization potential calculations, and the size effect of the metal cluster. In this chapter, attempts are made to investigate these effects using a combination of theoretical considerations, which are accompanied by the results of quantum calculations.

5.1 Effect of Madelung Charge

In the local reaction center (LRC) theory proposed by Anderson [6, 9, 10, 13, 25, 55], a point charge of -0.5 e is placed 10 Å away from the hydronium ion to represent the net effect of positive and negative ions present in the solution (Figure 5.1). The Madelung charge is used to represent a polarized electrolyte in an average and simple way in the simulations of electron transfer. Here, we study the role of this point charge, both its magnitude and its position with respect to the molecule, in the energetic calculations [48]. In addition, it will be shown that the change in energies

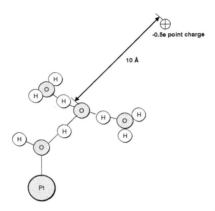

Figure 5.1: Position and magnitude of the Madelung charge that is used in the LRC theory to represent the effect of ions in the solution.

due to a point charge that is located at distances far away from the nuclei, can be predicted accurately by simple electrostatic calculations based on the Mulliken charge distributions in the molecule. We focus here on the oxidation reaction of $OH_{(ads)}$ in the chemical mechanism of PEM fuel cells as a typical example of an ET reaction. This reaction can be represented by a PtOH molecule as the computational system according to

$$PtOH \rightarrow PtOH^+ + e^-$$ (5.1)

The ionization potential (IP) is calculated as the difference between the electronic energy of the neutral and positive ions

$$IP_{PtOH} = E_{PtOH^+} - E_{PtOH}$$ (5.2)

The quantum computations were performed at the B3LYP level of theory with LANL2DZ basis set for Pt, and 6-31G** for O and H. The structure of the molecule was first

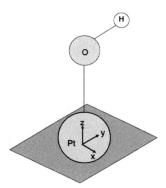

Figure 5.2: Coordinate system used to plot the change in the IP of PtOH in Figure 5.3.

optimized, and consequently the partial atomic charges (PACs) were calculated by the Mulliken population analysis. The optimized bond lengths were found to be: Pt-O=1.873 Å, O-H=0.976 Å, and bond angle: Pt-O-H=106.9°. The positive molecule was simulated at the optimized structure for the neutral molecule, but charged to form a cation. Then, according to equation (5.2), the IP was calculated and found to be 10.107 eV. The same procedure was followed when a point charge was also present in the computations.

Effect of In-plane Position. Figure 5.3 illustrates the IP of the neutral molecule as a function of x-y, which are the coordinates of the point charge $+0.5\ e$ with respect to the Pt atom at the origin (Figure 5.2). One observes a non-linear change of about 1 eV in the IP as the point charge moves in a 2Å×2Å square. The directional dependence of these variations can be attributed to the spatial shape of molecular orbitals for PtOH. For instance, the largest change is observed on the directions parallel to, and that perpendicular to the OH bond. It can be concluded here that if a point charge is close enough to the nuclei, the IP of the ion might undergo large variations

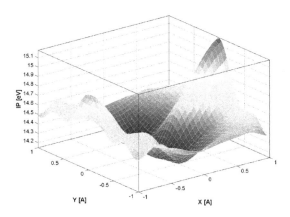

Figure 5.3: IP of PtOH on the left as affected by a $+0.5e$ point charge that is placed on the x-y plane of the molecule depicted in Figure 5.2.

depending on the orientation of the molecular orbitals. To see how close is close enough, the effect of distance is elaborated next.

Effect of Distance. The high directional interaction between the orbitals and a point charge is quite reasonable, when the charge is placed at a close distance to the nuclei. One expects less directional and more point-wise interaction as the orbital-charge distances are increased. To investigate this prediction, the point charge was moved away from the Pt atom on the Pt-O line from 2Å up to 10Å at 1Å steps. At each step, the IP of the molecule was calculated according to

$$IP_{system} = E_{PtOH^+ \& \text{ point charge}} - E_{PtOH \& \text{point charge}} \qquad (5.3)$$

On the other hand, the change in IP can also be estimated using the Coulombic interactions between the Mulliken charge distributions in the system and the point charge. This approximation is applied by dividing the electronic energy of the system

(= ion + point charge) into two parts: molecular and electrostatic.

$$E_{\text{ion \& point charge}} = E_{\text{molecular}} + E_{\text{electrostatic}} \tag{5.4}$$

Using equation (5.4) for PtOH and PtOH$^+$ in (5.3), one finds

$$\begin{aligned} IP_{\text{system}} &= \left(E_{\text{molecular}}^{\text{PtOH}^+} - E_{\text{molecular}}^{\text{PtOH}} \right) + \left(E_{\text{electrostatic}}^{\text{PtOH}^+} - E_{\text{electrostatic}}^{\text{PtOH}} \right) \\ &= IP_{\text{molecular}}^{\text{PtOH}} + \left(E_{\text{electrostatic}}^{\text{PtOH}^+} - E_{\text{electrostatic}}^{\text{PtOH}} \right) \end{aligned} \tag{5.5}$$

$E_{\text{molecular}}$ is obtained entirely by quantum calculations for a system without any point charge. Two quantum simulations for the neutral and positive molecule, without considering the point charge, give $IP_{\text{molecular}}^{\text{PtOH}}$. $E_{\text{electrostatic}}$ is calculated analytically from the interaction of the assumed point charge q_x with the PACs in the system according to the Coulomb law

$$E_{\text{electrostatic}} = \sum_{i=\text{Pt,O,H}} \frac{q_x q_i}{4\pi\epsilon_0\, r_{ix}} \tag{5.6}$$

where q_i is the PAC attributed to atom i, and r_{ix} is the distance of this atom to the point charge (Figure 5.4). Therefore, IP_{system} in equation (5.5) for a molecule in the presence of a point charge is be given by

$$IP_{\text{system}} = IP_{\text{molecular}}^{\text{PtOH}} + \frac{q_x}{4\pi\epsilon_0} \sum_{i=\text{Pt,O,H}} \frac{1}{r_{ix}} \left(q_i^{\text{PtOH}^+} - q_i^{\text{PtOH}} \right) \tag{5.7}$$

where $q_i^{\text{PtOH}^+}$ and q_i^{PtOH} are respectively the Mulliken charges of atom i in PtOH$^+$ and PtOH ions. Either equation (5.3) or (5.7) can be used to calculate IP_{system}, however, in the former, the point charge is practically present in the quantum calculations, while in the latter its effect is estimated through its interaction with the atomic charges. Figure 5.5 shows the IP of PtOH for these two methods as a function of the Pt-x distance. The direct quantum simulations, identified by markers, are compared with the curve of the analytical estimations as given by equation (5.7). When the point charge is more than 4Å away from the Pt atom, the approximation of

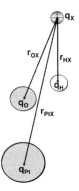

Figure 5.4: Electrostatic interactions between a point charge and a PtOH molecule as computed by the Coulomb formula.

equation (5.7) is quite successful in predicting the IP. In simple terms, when the point charge moves farther away from the molecule, it starts to see the molecular charge distribution as simple charges gathered on individual nuclei instead of 3D volumetric charges spread in space. This phenomenological behavior removes the need to perform further quantum calculations with the point charge explicitly present when Pt-x distances are large, say greater than 4Å for this system.

Effect of Charge Magnitude. By now, a +0.5 e point charge was used in all the simulations performed here. Depending on the nature of the solution and its activity, a negative and/or larger value might be applied. To investigate the effect of charge magnitude, its value was varied from -5.0 to +5.0 at 1.0 e steps, but its position was fixed at 10 Å. Since a -0.5e corresponds to an acid environment of 0.1M [55], the above range can reasonably represent the properties of a wide range of solutions within the assumptions of using a Madelung charge.

Figure 5.6 illustrates the IP of PtOH as a function of the magnitude of the point

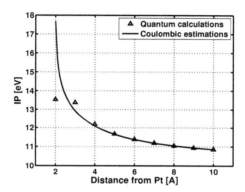

Figure 5.5: Comparison of the quantum-computed IP with the analytical estimations as a function of the ion-charge distance.

charge. The markers denote the direct quantum simulations, while the line is the estimation from equation (5.7). A perfect match is observed between the two series of values. It can thus be concluded that at large ion-charge distances, the IP of an ion is predictable by analytical estimations without the explicit presence of the point charge in quantum calculations.

How can the effect of a point charge, at large distances away from the ion, be calculated without resorting to the quantum calculations which determine the charge distribution in the 3D space? If the interaction between the point charge and electronic charge in the molecule as implemented in the quantum software can be simplified to the interactions between electrostatic charges, the point charge has not disturbed the PACs that have already been distributed around the nuclei in the absence of the point charge. This explanation can be verified by comparing the PACs in the neutral and positive molecule for different point charge magnitudes.

Figure 5.7 monitors the change in the PACs of Pt, O and H as the magnitude of

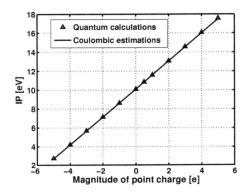

Figure 5.6: Comparison of the quantum-computed IP with the analytical estimations as a function of the charge magnitude when placed at 10Å away from the ions.

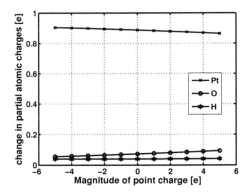

Figure 5.7: Change in the partial atomic charges between PtOH and PtOH$^+$ in terms of the magnitude of a point charge placed at 10Å away from the ions.

the point charge is varied. The zero value for the charge magnitude corresponds to the perfectly unperturbed charge distribution in the system. The changes in atomic charges are almost unaffected by the point charge, and more or less equal to those at the zero value. This confirms the picture of unperturbed atomic charges, when the point charge is placed far away from the ions.

5.2 Accuracy of Ionization Potentials

By definition, the IP of a molecule is the energy required to remove one mole of electrons from one mole of the material under study. EA is, on the other hand, the energy that is released by one mole of the material upon receiving one mole of electrons. The computational approach in the literature is similar to, but not exactly the same as the theoretical definition, rather they are closer to the definition of work function. Work function is the energy required to remove only one electron from a system. Similarly in the literature, IP or EA are estimated as the change in energy when only one electron is removed from or added to the system. The computational procedure of IP and EA has thus a more localized nature compared to their theoretical definitions, which reflect bulk properties. Despite the nomenclature in the literature not being completely correct, we continue to use the same terminology to avoid confusion.

In addition to inherent differences between theoretical definitions and computational implementations, no quantum computation leads to an absolutely exact value for the system energy. Even if the local reaction center (LRC) theory was a perfect theory for ET processes, its computational implementation would not reproduce precise results.

In the LRC theory, the ionization potential (IP) or electron affinity (EA) of the system is calculated in order to be compared with the work function of the electrode to assess if the electron transfer (ET) condition (4.1) is satisfied or not. The work function of the electrode in turn depends on the potential U of the electrode. Because the activation energies for ET reactions are reported in terms of U, the values of IP and EA have to be calculated as accurately as possible.

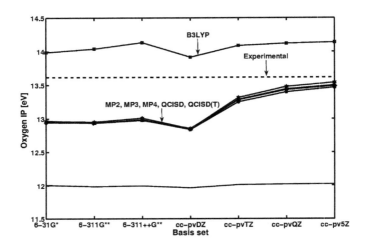

Figure 5.8: Comparison of the ionization potential of an oxygen atom obtained by different computational methods and basis sets with the experimental measurements.

In this section, we investigate the accuracy of these values computed by different methods and basis sets. We present results for the IP of a single O atom using HF, B3LYP, MP2, MP3, MP4, QCISD and QCISDT methods, and 6-31G**, 6-311G**, 6-311G++**, cc-pvDZ, cc-pvTZ, cc-pvQZ, and cc-pv5Z basis sets. It is expected that the results show improvement as better methods and better basis sets are used.

Figure 5.8 demonstrates the IP of an oxygen atom obtained by quantum computations, and also the experimental value. The HF method does not show any significant change for different basis sets, but remains about 1.6 eV below the experimental IP. The results by the B3LYP level of theory are also more or less indifferent with respect to the basis set. These values are on average about 0.4 eV higher than $IP_{experiment}$. All the other methods underestimate the IP. Both HF and B3LYP seem to generate, independent of the basis set, IPs that are accordingly lower and higher than the exact

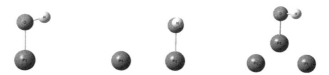

Figure 5.9: Ionization potential calculation of Pt_nOH molecules for $n = 1, 2, 3$ (left) Pt_1OH, (middle) Pt_2OH, (right) Pt_3OH.

value. This observation has been reported in the literature too [85, 88, 89].

MPn and Quadratic Configuration Interaction methods generate results very close to each other, and approach the exact value systematically as cc-pv$n$Z basis sets are applied. The value of IP calculated by QCISD(T)/cc-pv5Z, which are the best method and basis set, still underestimate the IP by 0.1 eV. Therefore, the quantum calculations of ionization potentials may deviate from the experimental values at least by a few tenths of electron volts. Another major factor that significantly influences the IP and EA of a molecule is the number of metal atoms, here Pt atoms, in the system. This issue will be explored in the next section.

5.3   Size Effect of Pt Cluster

The number of Pt atoms directly affects the calculation of ionization potentials since in the implementation of the LRC theory only one electron is removed from the whole system. The more Pt atoms are present in the system, the closer would be the IP of the whole system to the work function of platinum, which is about 6.35 eV. To show this saturation-like behavior, the IP of Pt_n-OH with $n = 1, 2, 3, 5, 7, 10$ (Figures 5.9-5.11) and Pt_n are computed at the B3LYP level using 6-31G* and LANL2DZ basis sets. The results are shown in Figure 5.12, and compared with the experimental work function of platinum. Clearly, as n increases, the IPs of both systems reduce toward the work function of Pt.

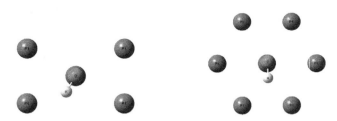

Figure 5.10: Ionization potential calculation of Pt_nOH molecules for $n = 5, 7$ (left) Pt_5OH, (right) Pt_7OH.

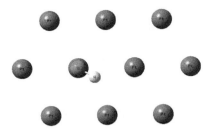

Figure 5.11: Ionization potential calculation of $Pt_{10}OH$ molecule.

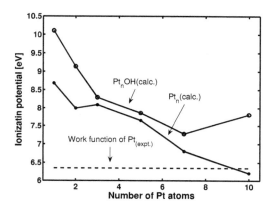

Figure 5.12: IP of two molecules Pt_n-OH and Pt_n as a function of the number of platinum atoms n.

The outer electronic shell of each Pt atom contains 10 electrons compared to 4 for an O atom, and 1 for an H atom. Thus, the electronic properties of Pt atoms soon dominate those of the whole molecule as two or more Pt's are present, even though there might be many non-Pt atoms in the system. As the number of metal atoms increases, the coverage of surface adsorbates decreases and consequently the properties of the computational model approaches those of the metal atoms. The domination of Pt over molecular properties can initiate serious issues in applying the LRC theory, in which one calculates ionization potentials primarily to characterize the adsorbates, and not the catalyst surface.

In addition to the total value of a system IP, its sensitivity to the removal of one electron is also affected by the number of Pt atoms. The number of valence electrons dramatically increases as a larger Pt cluster is used, for example 100 electrons for a Pt_{10} cluster. The current estimation of IP for a molecular system having 100 valence electrons would obviously be almost the same as the IP of the same system having

99 electrons.

Recall that the change in the structure of the reaction center in the LRC theory is exploited to adjust ionization potentials, and hence is compared with the electrode work function. If the molecule becomes indifferent to the removal of one electron, as in a large Pt cluster, the geometry of the reaction center has to undergo large variations in order to satisfy the electron transfer condition. Even if such variations could actually help in realizing the electron transfer condition, the structure itself may encounter unexpected distortions and represent a physically unacceptable configuration. In such cases, the validity of the results is questionable.

The LRC theory and/or its computational implementation at their current form require to be improved such that they can be used for systems with many atoms from the electrode surface, while still satisfactorily representing the properties of the adsorbates. For instance, one may consider removing more than one electron from a large system not only to approach the IP of the adsorbates, but also to keep the system sensitivity close to that of the adsorbates. However, this approach may lead to the removal of many electrons from a system, and may thereby tremendously affect the initial potential energy surface. Before the actual application of such an, or any other, extension to the LRC theory, they have to be thoroughly investigated and quantitatively characterized to enable interpreting the results in a more relevant way to physical expectations and experimental measurements. The reader is referred to Appendix A, which is focused on two possible pathways to improve the LRC theory.

5.4   Summary

In this chapter, we investigated a few aspects of the computational implementation of the LRC theory, which is based on quantum calculations to determine the activation energies of electron transfer reactions. It was shown that the effect of a Madelung charge can depend on the spatial orientation of molecular orbitals, if it is placed close enough to the molecule. On the other hand, a Madelung charge that is far away from a molecule does not affect the electronic charge distribution in the system. Hence its effect on energetics and ionization potentials can be accurately estimated by calculating the electrostatic attractions and repulsions between the point charge and the partial atomic charges collected on individual nuclei. Partial atomic charges can be evaluated once for all charge magnitudes by the Mulliken population analysis.

The accuracy of ionization potentials computed by quantum calculations suggested that HF and B3LYP might not provide improved results for higher quality basis sets. Methods based on perturbation and configuration interaction showed systematic improvements in the results compared to the experimental data when superior basis sets were used.

Finally, the effect of the number of platinum atoms in a system was studied. It was shown that the ionization potential of a system quickly approaches the work function of Pt as their number increases due to the enlargement of the electrode surface. This domination effect cannot be directly removed from quantum calculations as the electrons are not usually present in localized states. The cluster effect thus limits the the LRC theory in its current form to small systems. In the next chapter, we develop a particular computational methodology that circumvents many of possible issues in applying the LRC theory to systems which use a large Pt cluster. It will be shown that it is required to use reasonably large surfaces for determining ground states, but also to reduce the dominating effect of Pt on the properties of the reaction center in the activation energy calculations.

Chapter 6

Activation Energies of ET Reactions

Understanding the chemical phenomena involved in Proton Exchange Membrane (PEM) fuel cells requires a detailed knowledge about the important reactions at the atomic level. To go beyond a single-step representation of the electrochemistry in a PEM fuel cell, a suitable reaction mechanism was presented in Chapter 3 which includes both chemical and electrochemical (electron transfer) steps. The two main aspects of the present study are to establish a reaction mechanism and to estimate kinetic rate data. In this chapter, we study and compute the activation energies of three elementary electrochemical reactions

$$OH_{(ads)} + H^+ + e^- \rightleftharpoons H_2O \tag{6.1}$$

$$O_{(ads)} + H^+ + e^- \rightleftharpoons OH_{(ads)} \tag{6.2}$$

$$O_{2(ads)} + H^+ + e^- \rightleftharpoons O_2H_{(ads)} \tag{6.3}$$

and three electrochemical reactions involving OH- and O-interaction

$$OH_{(ads)} - O_{(ads)} + H^+ + e^- \rightleftharpoons OH_{(ads)} - OH_{(ads)} \tag{6.4}$$

$$OH_{(ads)} - OH_{(ads)} + H^+ + e^- \rightleftharpoons OH_{(ads)} - OH_{2(ads)} \tag{6.5}$$

116

$$O_{ads} + O_{ads} + H^+ + e^- \rightleftharpoons O_{ads} + OH_{ads} \qquad (6.6)$$

In the simulation of surface processes using a chemical mechanism, it is of paramount importance to utilize a set of consistent kinetic rate data. A consistent set of data can guide the simulations to predict correct physical trends even if the absolute rate values are not accurate enough. In the following section, such a uniform computational scheme is presented which is used throughout all quantum calculations performed in this chapter. After presenting the results for reactions (6.1) to (6.6), the theory of interaction in ET reactions is developed from a thermodynamic point of view. At the end of the chapter, it is shown that including interactions in the water discharge mechanism predicts the coverage of surface species in a very good agreement with experimental measurements.

6.1 Consistent Computational Scheme

The clusters that are studied here contain many atoms of hydrogen, oxygen, and platinum (Figure 6.1). In Figure 6.1, angles T_a and T_b are the bond angles between the adsorbed groups and the Pt surface. If only one Pt atom exists in the molecule, a value of 90 degrees can be used for bond angle T_a (Figure 6.1(a)). If the surface of the metal is represented by more than one Pt atom (Figures 6.1(b) and 6.1(c)), the bond angles should be relaxed in order to find the ground state of the system.

The computations show that the relaxation of surface bond angles, when dealing with two Pt atoms, can lead to unphysical configurations. For instance, Figure 6.2(a) demonstrates the initial structure of Pt-Pt.OH + H$^+$ at the triplet spin state. Figure 6.2(b) is the same precursor after several steps of optimization. The bond angle T_a has increased to more than 160 degrees after relaxation. Consequently, the structure is distorted with the hydronium ion penetrating the assumed surface of Pt. This unphysical situation is not observed for the optimization of the singlet spin state. Since the triplet state is the ground spin state of the system, a new technique must be adopted to prevent similar distortions.

In principle, one can either fix the bond angles or add more Pt atoms to prevent such structural distortions. If some bond angles have to be fixed, they must be fixed at their optimized values. However, as explained earlier, the inherent limitations of the local reaction center theory, which is used in the calculations of potential-dependent activation energy (PDAE) values, prohibit using a large number of Pt atoms. On the other hand, when using a smaller Pt cluster, the relaxation of surface angles may allow one or more of these angles to attain unphysical values through the optimization process. Therefore, the auxiliary Pt atoms must be used only in the initial optimization of the structure, but should not participate in PDAE calculations.

Based on the above arguments, a multi-step consistent methodology is adopted to determine the ground state of the reactants and products in the electrochemical reactions. First, additional Pt atoms are added to the free side of any adsorbent, and the structures are optimized for different states of spin. The ground spin state is determined from the results of this step in each case (Figure 6.3). Next, the extra Pt atoms are removed from the system, and the surface angles are fixed. We fix these angles at their averaged values corresponding to the ground state of reactants and products. This averaging technique is necessary to obtain the ground state structure of products starting from the ground state structure of reactants, and vice versa.

The degrees of freedom (DOF) that are varied in this initial step and in the constrained variation calculations are chosen to reflect the physics of the problem, such as the effects of interaction and/or solvation, without allowing the distortion of the structure. In general, these DOFs include the bond lengths within the hydronium ion, the bond lengths between the adsorbent atoms and the Pt sites, and those between the adsorbents themselves. Most of the bond angles, as well as the dihedral angles in the molecule, are fixed at their optimized values to prevent possible distortions. A point charge of -0.5 e is also added to the Hamiltonian of the system to account for the Madelung effect [55]. The computations performed in this step determine a set of consistent ground state configurations and energies for all reactions considered in this work. The calculation steps are elucidated in more details in the following:

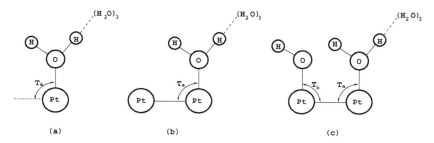

Figure 6.1: Definition of surface angles and their importance for optimization of systems with two Pt atoms.

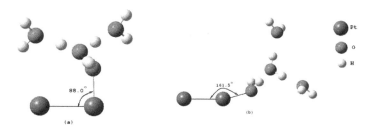

Figure 6.2: Relaxation of the surface angle in Pt-Pt.OH$_2$ takes the molecule from the initial structure (a) to the unphysical configuration (b).

1. *Ground state calculations.* In this step, the optimized structure of reactants is obtained. This is done for both oxidation and reduction reactions in a consistent way.

   (a) Add extra Pt atoms to the system, and optimize the structure of both neutral and positive molecule,

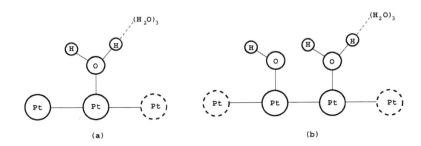

Figure 6.3: Auxiliary Pt atoms (represented by dashed lines) are added to prevent structural distortions.

(b) Average the surface angles found in these calculations,

(c) Remove the extra Pt atoms and re-optimize neutral and positive molecules using the averaged surface angles,

(d) Calculate the position of the point charge (Madelung sum) using the optimized structure,

(e) Optimize the structure again in the presence of the point charge.

Separate calculations for different multiplicities are usually required to determine the ground state of spin for each molecule. The averaged surface angles are then obtained from the multiplicities with the lowest energy. Normally, the addition of the point charge to the system does not appreciably affect the optimized geometry obtained before. Otherwise, one can adjust the location of the point charge with respect to the updated position of atoms, and iterate the optimization process. Finally, two optimized geometries and their corresponding energies obtained in this manner constitute the ground states of reactants for oxidation/reduction steps respectively.

2. *IP and EA calculations.* The next step is to find the ionization potential (IP) of the neutral molecule for the oxidation reaction, and the electron affinity (EA) of the positive molecule for the reduction reaction. These two quantities given

in Volts with respect to the hydrogen electrode are equal to the electrode potential at the ground states. The activation curves versus the electrode potential start from these values. To compute IP and EA, two single point quantum calculations are performed at the optimized structures that were found previously. Only the charge and multiplicity of these product molecules are different from those of the reactants. The product molecule for the neutral reactant is positively charged, and the product of the positive reactant is neutral. Having finished these calculations, the IP and EA are estimated as the change in the electronic energy of the system in losing and gaining an electron respectively.

3. *Activation energy curves.* In the final step, the input files for PDAE calculations are prepared, and the CICV method (Sec. 4.2) is used to obtain the activation curves.

The basis sets LANL2DZ+f were used for Pt, and CC-pVDZ for H and O. The f-function in the basis set for Pt was optimized to give the best binding energy of Pt-O in comparison with the high level method of Configuration Interactions [107].

6.2 $OH_{(ads)} \rightleftharpoons H_2O$ with and without OH-Interaction

Figure 6.4 shows the optimized structure of precursors for reactions (6.1) and (6.5). Note that the present computational methodology has effectively removed the possibility of structural distortions. Tables 6.1 and 6.2 give the important properties of the reactants and the products at their ground state. The row ψ_0 corresponds to the ionization (IP) of the reactants or the electron affinity (EA) of the products. The presence of an adjacent OH group has caused a decrease in the IP of the molecule by about 0.2 eV, and an increase in the EA by about 0.1 eV.

The degrees of freedom that are varied in the PDAE calculations are also shown in the same tables. The maximum change as a result of ET occurs in the bond lengths O_a-H_a and H_a-O_b. In going from the neutral to the positive energy surface of Pt-Pt.OH_2, the former bond length is increased by about 0.5 Å, while the latter one is decreased by approximately the same amount. The change in the bond length O_a-H_a in the Pt.OH-Pt.OH$_2$ case has decreased to about 0.4 Å due to the existence of the neighboring OH group.

The results of the PDAE calculations are given in Tables 6.3 and 6.4, and compared with the corresponding derived values obtained by the concept of microscopic reversibility in Figure 6.5. The good agreement between the curves shows the consistency of the PDAE calculations. As a result of this consistency, the reversible potential U_0 obtained from the intersection of PDAE curves is also close to the thermodynamical estimation of U_0 according to the relation [77]:

$$\Delta G = -nFU_0 \qquad (6.7)$$

where n is the number of exchanged electrons (equals one here), and F is the Faraday constant. The numerical results are compared in Table 6.5.

The slight corrugation in the activation energy curve of Pt-Pt.OH$_2$,shown in Figure 6.5 reduction reaction is due to the convergence problem in the quantum simulations for this reaction. Nevertheless, the difference between the thermodynamical value of U_0 and that of PDAE calculations, even in this case, is less than 0.02 V. The convergence problem is also observed in the oxidation reaction of Pt.OH-Pt.OH$_2$. As shown in Figure 6.5, when the activation energy attains a value of about 0.25 eV, the quantum calculations cannot find convergent solution anymore. A further investigation of the structure of the molecule reveals a distortion of the molecule with the $H_{a'}$ atom approaching O_d. This suggests the transition of $H_{a'}$ from O_a to O_d according to

$$\text{Pt.OH} - \text{Pt.OH}_2 \quad \rightarrow \quad \text{Pt.OH}_2 - \text{Pt.OH} \qquad (6.8)$$

We found the activation energy for this reaction to be 0.213 eV. The comparison of these activation energies clearly shows that at the electrode potential of around 1.1 V, where the activation energy of the ET reaches that of H-hopping, the latter reaction becomes more favorable.

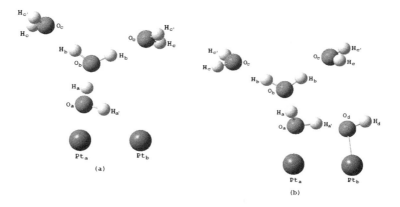

Figure 6.4: Optimized structure of Pt-Pt.OH$_2$ (a), and that of Pt.OH-Pt.OH$_2$ (b).

Figure 6.6 demonstrates the activation energies with and without the OH interaction. The dashed curves are for the base case of OH reducing to OH$_2$ and its reverse step. The solid curves are when the OH interaction is present. One observes a decrease of about 0.1 eV in both the oxidation and reduction steps due to this interaction. The reversible potential has increased by about 0.02 eV. In section 6.6, the thermodynamics of the relationship between interaction energies and the shift in reversible potentials is discussed in detail. The horizontal line in Figure 6.6 is the activation energy of hydrogen hopping according to reaction (6.8).

The activation energy for the hopping reaction is assumed to be independent of the electrode potential. The intersection point of the H-hopping with the oxidation curve is worth noting. If the electrode works below this potential, the activation energy of hopping becomes lower than that of the oxidation step and this happens because of

Table 6.1: Properties of the optimized reactants and products precursors in the oxidation/reduction reactions of Pt-Pt.OH$_2$.

Precursor / Properties	Pt-Pt.OH$_2$	Pt-Pt.OH$_2^+$
Energy [Hartree]	-543.9970648	-543.782726
ψ_0 [eV]	1.819	0.664
Pt$_a$-O$_a$	2.205	1.969
O$_a$-H$_a$	1.029	1.503
O$_a$-H$_{a'}$	0.9783	0.9854
H$_a$-O$_b$	1.538	1.044
O$_b$-H$_b$	0.9770	1.015
O$_b$-O$_c$	2.760	2.561
O$_c$-H$_c$	0.9679	0.9701
O$_c$-H$_{c'}$	0.9673	0.9695

the adjacent OH. Therefore, it is seen that OH at high coverage not only changes the activation energies but also initiates a new reaction, and makes H-hopping more favorable than the oxidation step. Changes in the reaction pathways due to interactions have been reported in the literature too. For example, Michaelides and Hu [68] found that the O-H bond dissociation in H$_2$O drops from 0.68 eV on a clean Pt surface to 0.33 eV on O-covered Pt. They concluded that the dissociation of H$_2$O through its reaction with O becomes much more favorable than its direct dissociation reaction. This result is in agreement with the experimental observations [40].

6.3 O$_{(ads)}$ ⇌ OH$_{(ads)}$ with and without OH-Interaction

For this system, first an extra Pt atom was added to the precursor Pt-Pt.OH[H$_2$O]$_3$ as explained in Sec. 6.1. This new system will here be denoted by Pt-Pt.OH[H$_2$O]$_3$-Pt. Starting from the initial structure in Figure 6.7(a), the simulations led to the distorted molecule shown in Figure 6.7(b), which is obviously not a physical configuration. To

Table 6.2: Properties of the optimized reactants and products precursors in the oxidation/reduction reactions of Pt.OH-Pt.OH$_2$.

Properties / Precursor	Pt.OH-Pt.OH$_2$	Pt.OH-Pt.OH$_2^+$
Energy [Hartree]	-619.8349614	-619.6198269
ψ_0 [eV]	1.623	0.712
Pt$_a$-O$_a$	2.069	1.938
Pt$_b$-O$_d$	1.951	1.925
O$_a$-H$_a$	1.020	1.392
O$_a$-H$_{a'}$	1.009	1.003
O$_d$-H$_d$	0.9689	0.9731
H$_a$-O$_b$	1.563	1.082
O$_b$-H$_b$	0.9762	1.008
O$_b$-O$_c$	2.766	2.581
O$_c$-H$_c$	0.9678	0.9698
O$_c$-H$_{c'}$	0.9672	0.9689

constrain the system in a more realistic way, a large Pt$_{10}$ cluster was used to simulate the Pt surface. Now, the molecule can be called Pt$_8$-Pt-Pt.OH[H$_2$O]$_3$, where the eight extra Pt atoms surround the initial Pt$_2$ cluster (Figure 6.8). From Figure 6.8(b), one can see that the optimized molecule is not distorted, and the relaxed surface angles are physically reasonable.

The energetics of different multiplicities are given in Table 6.6. It is seen that sextet and quintet are the ground states for the neutral and positive ion of Pt$_8$-Pt-Pt.OH[H$_2$O]$_3$, respectively. The sextet state energy is about 0.07 eV lower than that of the doublet and quartet, and 0.6 eV lower than the energy of the octet state. For the Pt$_8$-Pt-Pt.OH[H$_2$O]$_3^+$ ion, the quintet state energy is at least 0.27 eV lower than the energy of the singlet, triplet, and septet. While these differences in energetics are small with respect to the total energies, their effect on the rate data is appreciable.

Table 6.3:   Activation energies at various electrode potentials for the oxidation/reduction reactions of Pt-Pt.OH$_2$.

Pt-Pt.OH$_2$		Pt-Pt.OH$_2^+$	
Electrode potential [V]	E$_{act}$ [eV]	Electrode potential [V]	E$_{act}$ [eV]
0.7500	0.4943	0.6641	0
0.8500	0.4206	0.8000	0.0179
0.9500	0.3548	0.9000	0.0657
1.0500	0.2936	1.0000	0.1267
1.1500	0.2340	1.1000	0.1422
1.2500	0.1790	1.2000	0.1869
1.3500	0.1289	1.3000	0.2336
1.4500	0.0759	1.4000	0.2806
1.5500	0.0418	1.5000	0.3291
1.6500	0.0194	1.6000	0.4187
1.7500	0.0052	1.7000	0.5080
1.8187	0	1.8000	0.5683

Table 6.7 demonstrates averaging the four surface angles for the Pt$_8$-Pt-Pt.OH[H$_2$O]$_3$ molecule. Two bond angles T11 and T12, and two dihedral angles D11 and D12 are defined in the second column of the table. The atomic symbols and labels used in these definitions are given in Figure 6.8. For example, T11 is the bond angle between center number 11, i.e. O11, which is an oxygen atom and the two Pt atoms, Pt1 and Pt2. It is observed that the change in the bond angles T11 and T12 between the two optimized structures is about one degree. The dihedral angles D11 and D12, although relaxed, did not change during the optimization calculations.

In the next step, the extra Pt atoms are removed from the molecule and the surface angles are kept fixed at their averaged values (Figure 6.9). Note that the H[H$_2$O]$_3$ part of the molecule stands on no Pt atom. This is consistent with the simulation of the other electrochemical reactions, where the interaction of the hydronium ion with the surface is neglected.

Table 6.4: Activation energies at various electrode potentials for the oxidation/reduction reactions of Pt.OH-Pt.OH$_2$.

Pt.OH-Pt.OH$_2$		Pt.OH-Pt.OH$_2^+$	
Electrode potential [V]	E$_{act}$ [eV]	Electrode potential [V]	E$_{act}$ [eV]
1.0500	0.2529	0.7122	0
1.1500	0.1801	0.7500	0.0024
1.2500	0.1152	0.8500	0.0098
1.3500	0.0621	0.9500	0.0263
1.4500	0.0245	1.0500	0.0483
1.5500	0.0045	1.1500	0.0758
1.6254	0	1.2500	0.1111
-	-	1.3500	0.1580
-	-	1.4500	0.2204
-	-	1.5500	0.3001
-	-	1.6500	0.3967

Table 6.6 shows that doublet for the Pt-Pt.OH[H$_2$O]$_3$ precursor, and triplet for the Pt-Pt.OH[H$_2$O]$_3^+$ precursor are the ground states of spin. For the neutral molecule, the doublet state is more stable than the quartet state by about 0.2 eV. For the positive molecule, the triplet state is much more stable than the singlet by 1.43 eV, and by 0.91 eV with respect to the quintet.

A point charge of -0.5 e is then added to the Hamiltonian of the system to account for the acidic environment [5, 12]. The structure is then re-optimized and the ground state of the neutral system Pt-Pt.OH[H$_2$O]$_3$ and that of the positive system Pt-Pt.OH[H$_2$O]$_3^+$ are obtained. After this optimization step of the reactants, another two single point calculations are performed for the products as explained in Sec. 6.1. The IP of the neutral molecule, and EA of the positive one are then calculated from the energetics. The calculated values in Table 6.8 show that the activation curve for oxidation reaction starts from 2.11 V, and for reduction from 0.51 V. These potentials

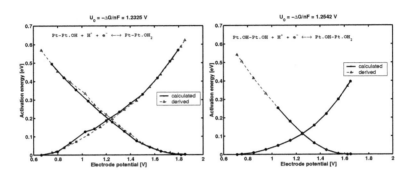

Figure 6.5: PDAE curves for Pt-Pt.OH$_2$ (a), and those of Pt.OH-Pt.OH$_2$ (b).

Table 6.5: Reversible potential U_0 for oxidation/reduction reactions of Pt-Pt.OH$_2$ and Pt.OH-Pt.OH$_2$.

Reaction \ U_0 [V] from:	Thermodynamics	PDAE calculations
Pt-Pt.OH$_2$	1.2325	1.2201
Pt.OH-Pt.OH$_2$	1.2542	1.2544

are with respect to the standard hydrogen electrode (SHE), which is at 4.6 V in the vacuum scale [78].

The degrees of freedom used in the PDAE calculations along with their optimized values are also given in Table 6.8. The bond length H6-O5 represents the distance between H$^+$ and the adsorbed oxygen O6. This bond length is about 0.5 Å shorter in the neutral molecule than in the positive molecule. On the other hand, the bond length O8-H6 is the distance between the H$^+$ and the three solvent molecules [H$_2$O]$_3$. In contrast to H6-O5, O8-H6 is approximately 0.6Å longer in the neutral than in the positive molecule. The switching of H$^+$ between the surface adsorbents and the

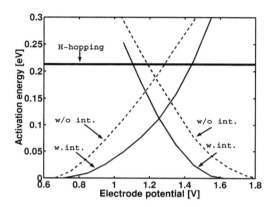

Figure 6.6: Activation curves for the oxidation/reduction steps of $OH_{(ads)}$ with and without interaction.

solvent molecules is the largest change among the degrees of freedom. This effect is also observed for other electrochemical reactions.

Activation curves for reaction (6.2) are given in Table 6.9 and plotted as a function of the electrode potential in Figure 6.10. As seen, the calculated activation energies are in a very good agreement with the derived values. The reversible potential given by the intersection of the oxidation curve with the reduction curve is 1.526 V. This quantity is equal to the thermodynamic estimation of the reversible potential, which is computed from the energies in Table 6.8.

Next, the results of calculations for reaction (6.4) are presented. In the first step, an extra Pt atom is added to the system to help finding the surface angle of the hydronium ion (Figure 6.11). Table 6.10 gives the energy values obtained for different states of spin. The singlet state for the neutral surface is by 0.114 eV more stable than the triplet state. The doublet state for the positive surface is 0.471 eV lower in energy than the quartet state. Using the optimized structure of the ground states

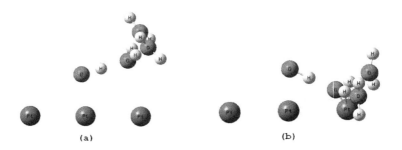

Figure 6.7: Optimization of Pt-Pt.OH[H$_2$O]$_3$-Pt from the initial structure shown in (a) led to the distortion of the geometry shown in (b).

of spin, four surface angles T5, T6, T19, and T20 are averaged in Table 6.11. The range of variations for all these angles is less than 5 degrees. In the next step, the extra Pt atom is removed from the system, and the surface angles are fixed at their average values. These frozen surface angles prevent the molecule of being distorted during the constrained variation simulations (Sec. 6.1).

Table 6.12 gives the energetics of the neutral molecule Pt.OH-Pt.OH[H$_2$O]$_3$ and the positive molecule Pt.OH-Pt.OH[H$_2$O]$_3^+$ for different spin states (Figure 6.12). It is seen that the singlet and the doublet states are more favorable than the triplet and the quartet states, which is consistent with the simulations on the extended cluster. At this stage of calculations, a Madelung charge of -0.5 e is added to the system, and the molecule is re-optimized. Table 6.13 shows the properties of the ground state for the two molecules. Here, the optimized values of the nine bond lengths are reported. These geometrical parameters are allowed to vary in the constrained variation calculations. Finally, the potential-dependent activation energy (PDAE) computations are performed and tabulated in Table 6.14. Figure 6.13 illustrates the calculated activation energy curves of reaction (6.4). The derived values are also plotted on this figure. The reversible potential at the intersection of the curves is in excellent agreement with its thermodynamic estimation, which is at about 1.74 V. The activation

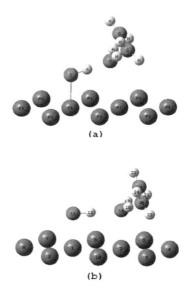

Figure 6.8: Optimization of Pt_8-Pt-Pt.OH$[H_2O]_3$ and averaging the surface angles: (a) initial structure, (b) optimized structure with labels used to define the geometry (see Table 6.7).

energy at this potential is approximately 0.27 eV.

To explicitly demonstrate the energetic interaction effects on ET reactions, the activation curves for reactions (6.2) and (6.4) are plotted together on Figure 6.14. It is observed that the reversible potential of the interacting case has moved by 0.2 V toward more positive potentials with respect to the non-interacting case. The activation energy of the reduction step has decreased by about 0.04 eV, while that of the oxidation step has increased by approximately 0.15 eV at every electrode potential. This means that the OH interaction strongly prevents the electrochemical dissociation of OH, and it makes the reverse step slightly more favorable.

Table 6.6: Energetics of $OH[H_2O]_3$ on Pt_{10} and Pt_2 clusters vs. the spin state. These systems are denoted by Pt_8-Pt-Pt.$OH[H_2O]_3$ and Pt-Pt.$OH[H_2O]_3$, and shown in Figure 6.8 and Figure 6.9, respectively.

Molecule	Multiplicity	Energy [Hartree]
Pt_8-Pt-Pt.$OH[H_2O]_3$		
	2	-762.4410038
	4	-762.4407037
	6†	-762.4436146
	8	-762.4207857
	10	-762.389638
	12	-762.3396038
Pt_8-Pt-Pt.$OH[H_2O]_3^+$		
	1	-762.1840128
	3	-762.2020191
	5†	-762.2120603
	7	-762.203007
Pt-Pt.$OH[H_2O]_3$		
	2†	-543.363185
	4	-543.355518
Pt-Pt.$OH[H_2O]_3^+$		
	1	-543.0627479
	3†	-543.11565
	5	-543.0820771

† The ground state of spin.

Table 6.7: Averaged surface angles for Pt_8-Pt-Pt.OH$[H_2O]_3$ precursor obtained from the ground spin states (see Table 6.6).

Angle	Centers[a]	Pt_8-Pt-Pt.OH$[H_2O]_3$	Pt_8-Pt-Pt.OH$[H_2O]_3^+$	Average
T11	O11-Pt1-Pt2	91.0429	90.4357	90.7393
D11 [b]	O11-Pt1-Pt2-Pt3	90.0	90.0	90.0
T12	H12-O11-Pt1	96.2785	97.7946	97.0366
D12 [b]	H12-O11-Pt1-Pt2	0.0	0.0	0.0

[a] See Figure 6.8
[b] These angles were allowed to vary during optimization, but they did not change.

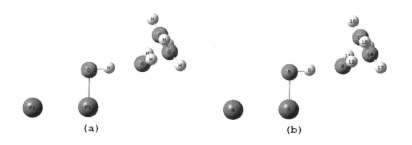

(a)   (b)

Figure 6.9: Optimization of Pt-Pt.OH$[H_2O]_3$ structure using the averaged surface angles: (a) atomic symbols, (b) labels used to define the geometry (see Table 6.8).

Table 6.8: Properties of the ground state for the Pt-Pt.OH[H$_2$O]$_3$ precursor (see Figure 6.9).

Property	Pt-Pt.OH[H$_2$O]$_3$	Pt-Pt.OH[H$_2$O]$_3^+$
Energy [Hartree]		
reactants	-543.3650039	-543.1398666
products	-543.118175	-543.3275852
ψ_0 [eV]	2.116	0.5081
Bond length in [Å]		
O5-Pt3	1.9067	1.8069
H6-O5	1.0058	1.5307
O8-H6	1.6626	1.0398
H12-O8	0.9753	1.0203
O14-O8	2.7755	2.5470
H18-O14	0.9677	0.9702
H17-O14	0.9673	0.9695

Table 6.9: Activation energies for $Pt\text{-}Pt.O + H^+ + e^- \rightleftharpoons Pt\text{-}Pt.OH$

Pt-Pt.OH $\rightarrow$ Pt-Pt.O + H$^+$ + e$^-$		Pt-Pt.O + H$^+$ + e$^-$ $\rightarrow$ Pt-Pt.OH	
Electrode potential [V]	E_{act} [eV]	Electrode potential [V]	E_{act} [eV]
0.7000	0.8561	0.5081	0
0.8000	0.7767	0.5000	0.0122
0.9000	0.6997	0.6000	0.0142
1.0000	0.6235	0.7000	0.0298
1.1000	0.5472	0.8000	0.0504
1.2000	0.4710	0.9000	0.0734
1.3000	0.3954	1.0000	0.0972
1.4000	0.3216	1.1000	0.1209
1.5000	0.2508	1.2000	0.1446
1.6000	0.1850	1.3000	0.1691
1.7000	0.1259	1.4000	0.1952
1.8000	0.0757	1.5000	0.2244
1.9000	0.0366	1.6000	0.2585
2.0000	0.0107	1.7000	0.2994
2.1000	0.0002	1.8000	0.3643
2.1166	0	-	-

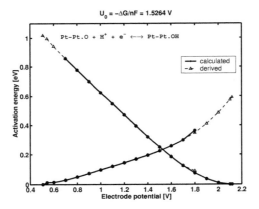

Figure 6.10: Potential-dependent activation energies for Pt-Pt.O + H$^+$ + e$^-$ ⇌ Pt-Pt.OH. The reversible potential at 1.526 V is consistent with its thermodynamic estimation.

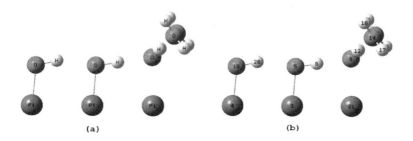

Figure 6.11: Optimization of Pt.OH-Pt.OH[H$_2$O]$_3$-Pt and averaging the surface angles: (a) molecular structure, (b) the same structure with the labels used to define the geometry (see Table 6.11).

Table 6.10: Energetics of OH-OH[H$_2$O]$_3$ on a Pt$_3$ cluster vs. the spin state. This system is denoted by Pt.OH-Pt.OH[H$_2$O]$_3$-Pt and shown in Figure 6.11.

Molecule	Multiplicity	Energy [Hartree]
Pt.OH-Pt.OH[H$_2$O]$_3$-Pt		
	1[†]	-738.3612064
	3	-738.3570092
Pt.OH-Pt.OH[H$_2$O]$_3$-Pt$^+$		
	2[†]	-738.1263602
	4	-738.1090552

[†] The ground state of spin.

Table 6.11: Averaged surface angles for Pt.OH-Pt.OH[H$_2$O]$_3$-Pt precursor obtained from the ground spin states (see Table 6.10).

Angle	Centers[a]	Pt.OH-Pt.OH[H$_2$O]$_3$-Pt	Pt.OH-Pt.OH[H$_2$O]$_3$-Pt$^+$	Average
T5	O5-Pt3-Pt4	94.5385	91.7890	93.1638
T6	H6-O5-Pt3	96.2184	93.5618	94.8901
T19	O19-Pt4-Pt3	92.3330	87.3243	89.8287
T20	H20-O19-Pt4	106.4448	109.9599	108.2023

[a] See Figure 6.11

Table 6.12: Energetics of OH-OH[H$_2$O]$_3$ on a Pt$_2$ cluster vs. the spin state. This system is denoted by Pt.OH-Pt.OH[H$_2$O]$_3$ and shown in Figure 6.12.

Molecule	Multiplicity	Energy [Hartree]
Pt.OH-Pt.OH[H$_2$O]$_3$	1[†]	-619.2025917
	3	-619.1957268[‡]
Pt.OH-Pt.OH[H$_2$O]$_3^+$	2[†]	-618.9494594
	4	-618.9487678

[†] The ground state of spin.
[‡] Energy of the last Self-Consistent Field (SCF) cycle. The system with this spin state failed to finish normally even after several iterations.

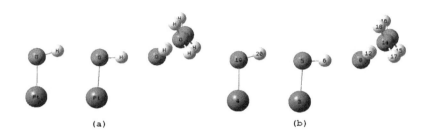

(a)   (b)

Figure 6.12: Optimization of Pt.OH-Pt.OH[H$_2$O]$_3$ and averaging the surface angles: (a) molecular structure, (b) the same structure with the labels used to define the geometry (see Table 6.13).

Table 6.13: Properties of the ground state for the Pt.OH-Pt.OH[H$_2$O]$_3$ precursor (see Figure 6.12).

Property	Pt.OH-Pt.OH[H$_2$O]$_3$	Pt.OH-Pt.OH[H$_2$O]$_3^+$
Energy [Hartree]		
reactants	-619.2035878	-618.9705735
products	-618.946937	-619.165996
ψ_0 [eV]	2.3839	0.7178
Bond length in [Å]		
O5-Pt3	1.9194	1.8116
H6-O5	1.0102	1.5852
O8-H6	1.6168	1.0248
H11-O8	0.9767	1.0252
O14-O8	2.7617	2.5345
H17-O14	0.9677	0.9704
H18-O14	0.9674	0.9698
O19-Pt4	1.8940	1.8900
H20-O19	0.9847	0.9853

Table 6.14: Activation energies for $Pt.OH\text{-}Pt.O + H^+ + e^- \rightleftharpoons Pt.OH\text{-}Pt.OH$

$Pt.OH\text{-}Pt.OH \rightarrow Pt.OH\text{-}Pt.O + H^+ + e^-$		$Pt.OH\text{-}Pt.O + H^+ + e^- \rightarrow Pt.OH\text{-}Pt.OH$	
Electrode potential [V]	E_{act} [eV]	Electrode potential [V]	E_{act} [eV]
1.0000	0.7918	0.7178	0.0000
1.1000	0.7165	0.8000	0.0211
1.2000	0.6445	0.9000	0.0320
1.3000	0.5741	1.0000	0.0512
1.4000	0.5043	1.1000	0.0759
1.5000	0.4347	1.2000	0.1038
1.6000	0.3655	1.3000	0.1334
1.7000	0.2973	1.4000	0.1636
1.8000	0.2313	1.5000	0.1940
1.9000	0.1693	1.6000	0.2247
2.0000	0.1132	1.7000	0.2566
2.1000	0.0655	1.8000	0.2906
2.2000	0.0288	1.9000	0.3285
2.3000	0.0060	2.0000	0.3724
2.3839	0.0000	2.1000	0.4247

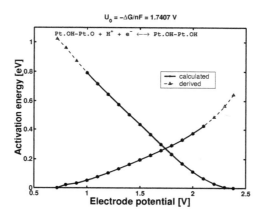

Figure 6.13: Potential-dependent activation energies for Pt.OH-Pt.O + H$^+$ + e$^-$ $\rightleftharpoons$ Pt.OH-Pt.OH. The reversible potential at 1.74 V is consistent with its thermodynamic estimation.

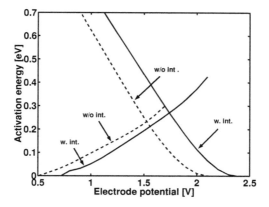

Figure 6.14: PDAE curves for the oxidation/reduction steps of O_{ads} with and without interaction.

6.4 $O_{(ads)} + H^+ + e^- \rightleftharpoons OH_{(ads)}$ with O-Interaction

In this section, activation energies for the interaction reaction

$$O_{(ads)} + O_{(ads)} + H^+ + e^- \rightleftharpoons O_{(ads)} + OH_{(ads)} \qquad (6.6)$$

are obtained.

Catalyst surface

First, we determined the dissociated configuration of $O_{2,(ads)}$ on a Pt cluster. A one-layer Pt_{13} cluster was used to find the optimized surface angles of the reactants and products. The initial bond length between the two $O_{(ads)}$ was increased to introduce their dissociation in the initial structure shown in Figure 6.15. The final optimized structure shows that the interacting O has moved toward a bridge site, while the O atom that reacts in the electron-transfer (ET) step has maintained its atop position. The final O-O distance at the ground state is approximately 4 Å, which is about 2 Å longer than the O-O bond length in reaction (6.3). This confirms that the subsequent computations of the ET activation energies occur while the two $O_{(ads)}$ are at their dissociated configuration.

Ground state of spin

To determine the ground state of spin, several calculations were performed with the results given in Table 6.15. For the neutral molecule, the quartet state has the lowest energy. This state is more stable by 1.5 eV than the doublet state, by 0.5 eV than the sextet, and by 0.3 eV than the octet state. Here it is note-worthy that the increase in energy as a function of spin after the ground state is not monotonic, that is the sextet is more unstable than the octet state. This complexity inherent in spin seems to be the reason for the computations of the sextet state to fail to converge with the required accuracy even after several iterations (Table 6.15).

The surface angles in the similar molecules, having different spins, are not very different from each other, because the coordinates of atoms are less sensitive to spin than

Figure 6.15: Optimization of surface angles for reaction $O_{(ads)} + O_{(ads)} + H^+ + e^- \rightleftharpoons$ $O_{(ads)} + OH_{(ads)}$: initial structure with atomic symbols, and final structure with labels to define the geometry in Table 6.16.

the energetics. For example, the bond angle H12-O11-Pt3 in Figure 6.15, also reported in Table 6.15, exhibits a deviation of about 2-3 degrees among different spins. Thus, the averaged surface angles, to be obtained in the next step, will not be very sensitive to using the geometry determined by any state of spin in this step. For the positive molecule, the quintet is the most stable state. Its energy is lower than the triplet by 0.5 eV, and by 0.17 eV than the septet state.

Averaging surface angles

Six surface angles were averaged among the values for the neutral and positively charged molecules. These are the bond angle and the dihedral angles of the two $O_{(ads)}$, and of the H^+ in the hydronium ion. As Table 6.16 shows, there is no or only a small variation in these angles between the neutral and the positive structures.

Table 6.15: Energetics of $O+OH[H_2O]_3$ on a Pt_{13} cluster vs. the spin state. This system is denoted by Pt_{11}-PtO-PtOH$[H_2O]_3$ and shown in Figure 6.15.

Molecule	Multiplicity	Energy [Hartree]	H12-O11-Pt3
Pt_{11}-PtO-PtOH$[H_2O]_3$			
	2	$-1011.50814257^{\ddagger}$	97.2
	$4^{\,\dagger}$	-1011.562813	97.1
	6	$-1011.54489384^{\ddagger}$	99.3
	8	-1011.5512238	96.1
	10	-1011.5300734	96.3
Pt_{11}-PtO-PtOH$[H_2O]_3^+$			
	1	-1011.2941024	97.6
	3	-1011.3091261	98.5
	$5^{\dagger}$	-1011.3280329	98.0
	7	-1011.3217025	98.1
	9	-1011.3080487	97.1

† The ground state of spin
‡ Energy of the last Self-Consistent Field (SCF) cycle

Table 6.16: Averaged surface angles for $O+OH[H_2O]_3$ on a Pt_{13} cluster obtained from the ground spin states (see Table 6.15).

Angle	Centers[a]	Pt_{11}-PtO-PtOH$[H_2O]_3$	Pt_{11}-PtO-PtOH$[H_2O]_3^+$	Average
T11	O11-Pt2-Pt4	89.6809	90.1520	89.9165
D11 [b]	O11-Pt2-Pt4-Pt3	90.0	90.0	90.0
T12	H12-O11-Pt2	97.1239	98.07661	97.6003
D12 [b]	H12-O11-Pt2-Pt4	180.0	180.0	180.0
T24	O22-Pt4-Pt2	129.1816	129.9146	129.5481
D24 [b]	O22-Pt4-Pt2-O11	0.0	0.0	0.0

[a] See Figure 6.15
[b] These angles were allowed to vary during optimization, but they did not change.

Ground state of reactants and products

Next, the catalyst surface represented by Pt_{13} in Figure 6.15 was reduced to a Pt_2 cluster as in Figure 6.16, and the surface angles were fixed. Once again, the structure was optimized for several states of spin. The energetics in Table 6.17 show that the doublet state for the neutral, and the quintet state for the positive molecule, are the ground states of spin. The doublet state is lower in energy by about 0.14 eV than the quartet. The quintet state is also much more stable than the singlet and the septet states by more than 1 eV. However, it does not exhibit such an appreciable stability over the triplet state being only 0.027 eV lower. This difference is very close to the thermal energy k_BT (=0.026 eV) at room temperature that is naturally introduced into the system through its contact with the environment.

Figure 6.16: Structure of the reaction center for $O_{(ads)} + O_{(ads)} + H^+ + e^- \rightleftharpoons O_{(ads)} -$ $OH_{(ads)}$: (left) the structure of the molecule with atomic symbols, (right) the molecule with labels to define the geometry in Table 6.17.

Moreover, since the ground state of the neutral molecule is doublet, one expects that the ground state of the positive molecule be either a singlet or a triplet state, and not a quintet. This expectation follows Hund's rules, which determine the order in which the orbitals are filled by electrons [20]. The electronic shell in a neutral molecule with a multiplicity of two includes a single unpaired electron. If the molecule is ionized and loses that electron, two cases are possible. If after ionization, the vacant

Table 6.17: Energetics of $O+OH[H_2O]_3$ on a Pt_2 cluster vs. the spin state. This system is denoted by $PtO-PtOH[H_2O]_3$, and is shown in Figure 6.16.

Molecule	Multiplicity	Energy [Hartree]
$PtO-PtOH[H_2O]_3$		
	$2^\dagger$	-618.559454
	4	-618.5544484
	6	-618.53598
$PtO-PtOH[H_2O]_3^+$		
	1	-618.2489181
	$3^\ddagger$	-618.2974594
	$5^{\dagger\,\ddagger}$	-618.2984354
	7	-618.2579207

$\dagger$ The calculated ground states of spin.
$\ddagger$ Two spin states that are very close in energy for the positive molecule.

orbital remains empty, the multiplicity equals to one. On the other hand, if according to Hund's rules that orbital becomes half occupied, the multiplicity will increase to three. Therefore, the observed jump of multiplicity of the neutral molecule from two to five upon ionization is not consistent with the rules about the ground state of a multi-electron system. Although these rules are not followed completely in all atoms and molecules, a deviation from them is an indicator of more complexity in determining the ground state of the electronic configuration.

Based on the above two arguments, the triplet was chosen, instead of the quintet, as the ground state of spin for the positive molecule. Nevertheless, whether the observed issues with the spin are purely computational artifacts, or they stem from real physical complications that are reflected in quantum computations, some difficulties in locating the transition states of the ET steps in reaction (6.6) can be anticipated. These difficulties did appear in the PDAE calculations for the reduction step as it is discussed below.

Table 6.18: Properties of the ground state for the $O_{(ads)}+OH[H_2O]_{3,(ads)}$ on a Pt_2 cluster (see Figure 6.16).

Property	PtO-PtOH[H$_2$O]$_3$	PtO-PtOH[H$_2$O]$_3^+$
Energy [Hartree]		
reactants	-618.5617747	-618.3220868
products	-618.3001484	-618.5207042
ψ_0 [eV]	2.5193	0.8047
Bond length in [Å]		
O5-Pt4	1.7475	1.7684
O6-Pt3	1.8990	1.8078
H7-O6	1.0130	1.5530
O9-H7	1.6196	1.0331
H13-O9	0.9764	1.0217
O15-O9	2.7640	2.5422
H18-O15	0.9677	0.9703
H19-O15	0.9674	0.9696

Table 6.19: ψ of the ground states for $O_{(ads)} + H^+ + e^- \rightleftharpoons OH_{(ads)}$ with and without interactions.

Reaction — Ground state IP & EA	IP [eV]	EA [eV]
$O_{(ads)} + O_{(ads)} + H^+ + e^- \rightleftharpoons O_{(ads)} + OH_{(ads)}$	2.5193	0.8047
$O_{(ads)} + H^+ + e^- \rightleftharpoons OH_{(ads)}$	2.116	0.5081
$OH_{(ads)} + O_{(ads)} + H^+ + e^- \rightleftharpoons OH_{(ads)} + OH_{(ads)}$	2.3839	0.7178

The reaction center for PDAE calculations is shown in Figure 6.16, and the properties are given in Table 6.18. The values of ψ at the ground states are equal to the ionization potential (IP), and electron affinity (EA) of the reaction center. These values have moved to more positive potentials with respect to the non-interaction case in reaction (6.2). For the ease of comparison, the corresponding IP and EA for reactions (6.2)-(6.3), and the O-reduction reaction with $OH_{(ads)}$-interaction in reaction (6.4) are given in Table 6.19.

Compared to the values given for reaction (6.2), the oxidation potential has increased by 0.4 eV, and the reduction potential by 0.3 eV. Since the structure of the hydronium ion and the reacting $O_{(ads)}$ have not appreciably changed, one concludes that the interacting $O_{(ads)}$ has increased the IP and EA of the whole system directly by being present, and not indirectly through changing the structure of the molecule. The first redox step of ORR in reaction (6.3) exhibits a different effect, where the IP is lower by about 0.5 eV, while the EA is only slightly higher than in reaction (6.2). The reason for this difference is the presence of a strong O-O bond in reaction (6.3), and the weakness or absence of such a bond in reaction (6.6). Comparing the effects of O-interaction with OH-interaction, one sees that their effects are more or less similar to each other.

In the final step of quantum simulations, the activation energies for reaction (6.6) were calculated. The calculations for the oxidation step were converged within acceptable tolerances. In contrast, almost all the corresponding calculations for the reduction steps failed to converge (Figure 6.17). It was found that the reason for this failure had been the corrugated energy surface of the positive molecule. This difficulty is assumed to be related to the observed issue with determining the correct state of spin, as it was extensively discussed above.

Hence instead of direct computation, we used the microreversibility principle to obtain the reduction curve from the oxidation curve. Based on this principle, at any given electrode potential, the structure of the transition states are the same for both the forward and backward steps. In this special case, since the transition states were known from the oxidation step, the activation energies for the reduction step were readily determined. In Figure 6.17, these are denoted by *derived* values. Note that the slope of the derived reduction curve at its lower extreme, that is the dotted line

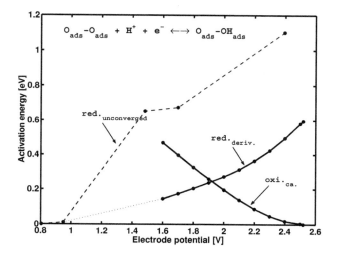

Figure 6.17: Potential-dependent activation energies for $O_{(ads)} + O_{(ads)} + H^+ + e^- \rightleftharpoons O_{(ads)} + OH_{(ads)}$. The oxidation curve (oxi.$_{ca.}$) was used to obtain the reduction activation energies denoted by red.$_{deriv.}$.

segment, is consistent with the slope that is given by the first two points on the calculated curve. This partial consistency can be interpreted as another support for the validity of the derived curve.

For further applications, we fitted a curve to the first two converged points on the reduction curve along with the derived points from the oxidation step. This fitted curve will be used to give the activation energies for the forward step in reaction (6.6).

The activation energies are given in Table 6.17, and are distinguished by appropriate labels for the reduction step. Figure 6.18 compares the activation energies for reactions (6.2) and (6.6). It is seen that the effect of an adjacent $O_{(ads)}$ on reaction (6.2) is very similar to the effect of an $OH_{(ads)}$ on the same reaction. Roughly speaking,

Table 6.20: Activation energies for $O_{(ads)}+OH[H_2O]_{3,(ads)}$ on a Pt_2.

$O_{(ads)}+OH_{(ads)} \rightarrow O_{(ads)}+O_{(ads)} + H^+ + e^-$		$O_{(ads)}+O_{(ads)} + H^+ + e^- \rightarrow O_{(ads)}+OH_{(ads)}$	
Electrode potential [V]	E_{act} [eV]	Electrode potential [V]	E_{act} [eV]
2.5193	0.0000*	2.500	0.5812†
2.500	0.0043*	2.400	0.4965†
2.400	0.0181*	2.300	0.4244†
2.300	0.0465*	2.200	0.3650†
2.200	0.0871*	2.100	0.3155†
2.100	0.1376*	2.000	0.2736†
2.000	0.1958*	1.900	0.2372†
1.900	0.2594*	1.800	0.2045†
1.800	0.3268*	1.700	0.1742†
1.700	0.3964*	1.600	0.1452†
1.600	0.4674*	1.500	0.1178‡
		1.400	0.0923‡
		1.300	0.0690‡
		1.200	0.0484‡
		1.100	0.0309‡
		1.000	0.0167‡
		0.8193	0.007 *
		0.8047	0.000 *

* Converged values calculated by quantum computations.
† These activation energies for the reduction step are the derived values using the oxidation curve and the ground state of the positive molecule.
‡ These activation energies for the reduction step are fitted values.

the activation energy for the reduction step has decreased by around 0.1 eV, while the oxidation curve has risen by approximately 0.3 eV. According to the theoretical relations that will be developed later in this chapter, equation (6.17) predicts a 0.4 V shift of the reversible potential to more positive values. The predicted shift is also observed in Figure 6.18.

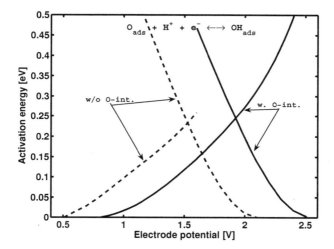

Figure 6.18: Comparison of PDAE curves for reaction $O_{(ads)} + H^+ + e^- \rightleftharpoons OH_{(ads)}$ with and without the interaction of an $O_{(ads)}$.

6.5 $O_{2(ads)} \rightleftharpoons O_2H_{(ads)}$

For this reaction, the precursor $Pt.O\text{-}Pt.OH[H_2O]_3$ was simulated by quantum computations. Figure 6.19 shows the atomic structure of the optimized molecule. No extra Pt atom was needed to constrain the surface angles of adsorbents since all the angles were physically reasonable. This feature can be related to the strong bond between the two adsorbed oxygen atoms. Table 6.21 gives the energetics of different multiplicities for this system. It is seen that the doublet and the singlet states are the ground states for the neutral and positive molecules respectively. The doublet state for the neutral molecule $Pt.O\text{-}Pt.OH[H_2O]_3$ is 0.62 eV more stable than the quartet state. The energy of the singlet state for $Pt.O\text{-}Pt.OH[H_2O]_3^+$ is 0.12 eV lower than the triplet energy, and about 0.92 eV lower than the quintet energy.

The bond angle and dihedral angle of the hydronium ion $H[H_2O]_3$ are averaged in Table 6.22. The bond angle T7 varies by less than one degree, while the range of changes in dihedral angle D7 is about ten degrees. Note that the hydronium ion is simulated as a non-adsorbed species. This is in agreement with the simulation of the other electrochemical reactions.

In the next step, the surface angles T7 and D7 are fixed at their average values, and a -0.5 e point charge is added to the system. The ground states of the reactants and products are then found by re-optimizing these new molecular systems. The results are shown in Table 6.23. The values denoted as ψ_0 give the starting potentials of activation curves with respect to SHE. Eight bond lengths and two bond angles defined in the table were allowed to vary in the constrained variation calculations. The bond lengths H7-O6 and O9-H7 show the largest change among the ten degrees of freedom. The H^+ atom is closer to the adsorbed oxygen by about 0.5 Å than to the three solvent molecules $[H_2O]_3$ in the neutral system. The reverse is true for the positive system, where the H^+ atom is now much closer to the three water molecules.

The results of PDAE calculations are given in Table 6.24, and plotted as a function of the electrode potential in Figure 6.20. It is observed that the derived and calculated curves are in excellent agreement. The reversible potential from the intersection of the two activation curves and the one computed from the thermodynamics are both 0.95 V. This potential requires a 0.141 eV activation energy for both directions of the reaction. These values are comparable with 0.139 eV at 0.92 V reported by Anderson *et al.* [8] using different basis sets.

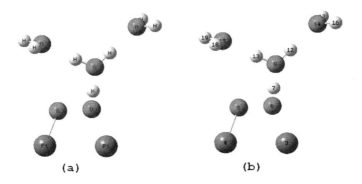

Figure 6.19: Optimization of Pt.O-Pt.OH[H$_2$O]$_3$ structure: (a) atomic symbols, (b) labels used to define the geometry (see Table 6.23)

Table 6.21: Energetics of Pt.O-Pt.OH[H$_2$O]$_3$ vs. the spin state.

Molecule	Multiplicity	Energy [Hartree]
Pt.O-Pt.OH[H$_2$O]$_3$		
	2[†]	-618.5260989
	4	-618.5031764
Pt.O-Pt.OH[H$_2$O]$_3^+$		
	1[†]	-618.3009744
	3	-618.2964609
	5	-618.267189814

[†] The ground state of spin.

Table 6.22: Averaged surface angles for Pt.O-Pt.OH[H$_2$O]$_3$ precursor obtained from the ground spin states (see Table 6.21, and Figure 6.19).

Angle	Centers[a]	Pt.O-Pt.OH[H$_2$O]$_3$	Pt.O-Pt.OH[H$_2$O]$_3^+$	Average
T7	H7-O6-O5	98.3478	97.8519	98.0998
D7	H7-O6-O5-Pt4	250.5720	239.6137	245.0934

[a] See Figure 6.19

Table 6.23: Properties of the ground state for Pt.O-Pt.OH[H$_2$O]$_3$ (see Figure 6.19).

Property	Pt.O-PtOH[H$_2$O]$_3$	Pt.O-PtOH[H$_2$O]$_3^+$
Energy [Hartree]		
reactants	-618.529172	-618.3251766
products	-618.300455	-618.5144919
ψ_0 [eV]	1.6238	0.5516
Bond lengths [Å] & bond angles [degrees]		
O5-Pt4	1.9010	1.9065
O5-Pt4-Pt3	74.5824	70.6125
O6-O5	1.5212	1.4642
O6-O5-Pt4	109.4530	109.8032
H7-O6	1.0558	1.5383
O9-H7	1.4412	1.0254
H12-O9	0.9790	1.0194
O14-O9	2.7351	2.5451
H16-O14	0.9678	0.9701
H17-O14	0.9674	0.9694

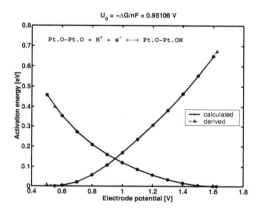

Figure 6.20: Potential-dependent activation energies for Pt.O-Pt.O + H$^+$ + e$^-$ $\rightleftharpoons$ Pt.O-Pt.OH. The reversible potential at 0.951 V is consistent with its thermodynamic estimation.

Table 6.24: Activation energies for Pt.O-Pt.O + H$^+$ + e$^-$ $\rightleftharpoons$ Pt.O-Pt.OH

Pt.O-Pt.OH → Pt.O-Pt.O + H$^+$ + e$^-$		Pt.O-Pt.O + H$^+$ + e$^-$ → Pt.O-Pt.OH	
Electrode potential [V]	E_{act} [eV]	Electrode potential [V]	E_{act} [eV]
0.5000	0.4578	-	-
0.6000	0.3533	0.5516	0
0.7000	0.2716	0.6000	0.0040
0.8000	0.2084	0.7000	0.0199
0.9000	0.1590	0.8000	0.0567
1.0000	0.1194	0.9000	0.1074
1.1000	0.0858	1.0000	0.1685
1.2000	0.0571	1.1000	0.2348
1.3000	0.0329	1.2000	0.3060
1.4000	0.0153	1.3000	0.3819
1.5000	0.0043	1.4000	0.4640
1.6000	0.0000	1.5000	0.5532
1.6238	0	1.6000	0.6488

6.6 Thermodynamics of Interaction Effects

In this section, we develop the thermodynamical formulation of interaction in electron-transfer (ET) reactions. The fundamentals of this model is essentially described in the Marcus theory [60] for electron-transfer processes. However, from a computational point of view, this formulation is expressed within the local reaction center (LRC) theory proposed by Anderson and Albu [6]. A comprehensive background on the LRC theory can be found in Refs. [16, 55]. This formulation enables us to check the thermodynamic consistency of ab-initio calculations for interactions, and also gain insight into the nature of interactions.

Energetics-Potential Relationship

Consider the energy curves for a system before and after an ET reaction, without (case I) and with (case II) the presence of interactions (Figure 6.21). The goal is to find a relationship between the changes in energetics and in reversible potentials. First, let us write the equations that define activation energies, at an arbitrary potential U of the electrode. For the initial system without interaction (case I):

$$\text{activation energy for case I in reduction step} = E_{\text{act,I}}^{+}(U) \triangleq E_{\text{A1}} - E_{\text{A0}} \qquad (6.9)$$

$$\text{activation energy for case I in oxidation step} = E_{\text{act,I}}^{0}(U) \triangleq E_{\text{A2}} - E_{\text{A3}} \qquad (6.10)$$

and for the system with interaction (case II):

$$\text{activation energy for case II in reduction step} = E_{\text{act,II}}^{+}(U) \triangleq E_{\text{B1}} - E_{\text{B0}} \qquad (6.11)$$

$$\text{activation energy for case II in oxidation step} = E_{\text{act,II}}^{0}(U) \triangleq E_{\text{B2}} - E_{\text{B3}} \qquad (6.12)$$

Because all the activation energies are computed at the same electrode potential, one can write the following relations within the LRC theory:

$$eU = E_{\text{A1}} - E_{\text{A2}} \qquad (6.13)$$

$$eU = E_{\text{B1}} - E_{\text{B2}} \qquad (6.14)$$

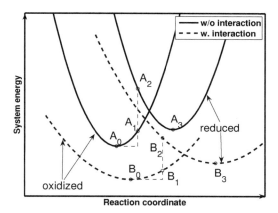

Figure 6.21: System energy without and with interaction effects as a function of reaction coordinate

where e is the electronic charge. The reversible potential for both cases is determined from thermodynamics [77]:

$$eU_{o,I} = E_{A0} - E_{A3} \tag{6.15}$$

$$eU_{o,II} = E_{B0} - E_{B3} \tag{6.16}$$

We are seeking a relation between the change in the activation energy for the oxidation step ΔE_{act}^0, in that of the reduction step ΔE_{act}^+, and the change in the reversible potential ΔU_o caused by adsorbate interactions. To this end, we start from equation (6.12)

$$
\begin{aligned}
E^0_{\text{act,II}} &\underset{(6.12)}{=} E_{\text{B2}} - E_{\text{B3}} \\
&\underset{(6.14)}{=} E_{\text{B1}} - eU - E_{\text{B3}} \\
&\underset{(6.11,6.13)}{=} (E^+_{\text{act,II}} + E_{\text{B0}}) - (E_{\text{A1}} - E_{\text{A2}}) - E_{\text{B3}}
\end{aligned}
$$

Substituting from equations (6.10,6.9) leads to

$$
\begin{aligned}
\therefore \underset{(6.10,6.9)}{} \Delta E^0_{\text{act}} &= \Delta E^+_{\text{act}} + (E_{\text{B0}} - E_{\text{B3}}) + (-E_{\text{A0}} + E_{\text{A3}}) \\
&\underset{(6.15,6.16)}{=} \Delta E^+_{\text{act}} + eU_{\text{o,II}} - eU_{\text{o,I}} \\
&= \Delta E^+_{\text{act}} + e\Delta U_o
\end{aligned}
$$

From this it follows

$$
\therefore \Delta E^0_{\text{act}} = \Delta E^+_{\text{act}} + e\Delta U_o \tag{6.17}
$$

This is the desired relationship, which states that the change in activation energy of the oxidation step at any electrode potential due to interactions should equal the sum of the change in the reduction step and the change in the reversible potential due to interactions. Since the change in the reversible potential ΔU_o only depends on the thermodynamics of the system, the difference between ΔE^0_{act} and ΔE^+_{act} should be independent of the electrode potential:

$$
\Delta E^0_{\text{act}} - \Delta E^+_{\text{act}} = \text{Constant} \tag{6.18}
$$

Repulsive Interaction

If the interaction between the reaction center (RC) and the interacting agent (IA) is repulsive, the electronic structure of the IA remains almost unchanged by removing one electron from the RC or when an electron is added to the oxidized RC. The electronic structure of the RC and its oxidized state RC^+ are not affected by the presence of the IA either, because no orbital is shared between the two species. This means that the ground states of the RC and RC^+ do not change due to the presence of an IA. As a result, the reversible potential will not be different or at most only slightly different from the initial system, *i.e.*:

$$\Delta U_o \simeq 0 \qquad (6.19)$$

From equation (6.17), one immediately concludes that the change in the activation energies for oxidation and reduction steps should be the same:

$$\Delta E^0_{act} \underset{\Delta U_o \to 0}{=} \Delta E^+_{act} \qquad (6.20)$$

An example of this case is the oxidation/reduction of $OH_{(ads)}$ in reaction (6.5). As Figure 6.6 shows, the reversible potential has moved only by 0.02 V due to the interaction effect. On the other hand, the activation energies for both steps have decreased by almost the same amount of about 0.1 eV. These changes in the reversible potential and in the activation energies are in agreement with the prediction made by equation (6.20). Based on this discussion, one can conclude that the OH interaction with ET steps of $OH_{(ads)}$ is of a repulsive nature.

Attractive Interaction

In this case, the electronic structure of the RC is affected by the presence of the IA. One can think that the highest occupied molecular orbital (HOMO) electron of the RC is now shared by the IA in a constructive way due to the attractive nature of the interaction. The ground state of the new system is more stable than the ground state of the initial system. Because the shared orbital reduces the system energy more than an empty orbital, the neutral state becomes more stable than the positive state. This means that the reversible potential has to shift toward more positive potentials:

$$\Delta U_\mathrm{o} > 0 \qquad (6.21)$$

Using equation (6.17) once again, one can conclude that the change in interaction energy of oxidation step is larger than (i.e. more positive than) the corresponding change in the reduction step

$$\Delta E_\mathrm{act}^0 \underset{\Delta U_\mathrm{o}>0}{>} \Delta E_\mathrm{act}^+ \qquad (6.22)$$

The oxidation/reduction of $O_\mathrm{(ads)}$ in reaction (6.4) is an example of this case. Figure 6.14 shows that the reversible potential has increased by a large value of about 0.2 V. The reduction activation energy has decreased only by 0.04 eV, while the oxidation activation energy has increased by 0.15. It is seen that these results are in agreement with equation (6.17), and inequality (6.22). These results suggest an attractive interaction between $OH_\mathrm{(ads)}$ and $O_\mathrm{(ads)}$.

Figure 6.22 illustrates the residual of equation (6.17) for the oxidation/reduction steps of $O_\mathrm{(ads)}$ and $OH_\mathrm{(ads)}$ interacting with an adjacent OH as a function of the electrode potential. The computational error for the $O_\mathrm{(ads)}$ reaction is almost zero except at one point. The corresponding error for the $OH_\mathrm{(ads)}$ reaction is approximately 0.015 eV. Considering the fact that any of these two curves is the superposition result of four series of PDAE calculations, each involving numerous quantum and constrained variation computations, the accuracy of the interaction simulations is remarkably good.

6.7   Application to Water Discharge

The water discharge mechanism is a subset of the oxygen reduction mechanism that includes the dissociation of water as its principle step, but excludes reactions involving

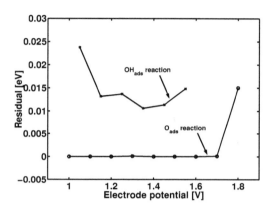

Figure 6.22: Residual of equation (6.17) for the OH interaction with the electrochemical reactions of $O_{(ads)}$ and $OH_{(ads)}$.

molecular oxygen. Studying the water discharge mechanism is a good validation for the overall chemistry of PEM fuel cells to demonstrate an application of the chemical kinetic rate data that were computed in this work [101]. The chemical mechanism consists of the following elementary steps

$$adsorption\ of\ water: \quad H_2O_{(aq)} \rightleftharpoons H_2O_{(ads)} \tag{6.23}$$

$$ET\ dissociation\ of\ water: \quad H_2O_{(ads)} \rightleftharpoons OH_{(aq)} + H^+ + e^- \tag{6.24}$$

$$ET\ dissociation\ of\ OH: \quad OH_{(ads)} \rightleftharpoons O_{(ads)} + H^+ + e^- \tag{6.25}$$

$$H-hopping: \quad OH_{(ads)} + OH_{(ads)} \rightleftharpoons O_{(ads)} + H_2O_{(ads)} \tag{6.26}$$

where the main step is (6.24). Figure 6.23 shows the coverage of O-containing species, $i.e.$ $O_{(ads)}$ and $OH_{(ads)}$, versus the electrode potential. The experimental curve has been extracted from Wang $et\ al.$ [108], while the other three curves are the results of Monte-Carlo simulations performed using the ET activation energies obtained in this chapter. The rate constants for non-ET reactions (6.23) and (6.26) are taken from the literature and from Ref. [107], respectively. Interaction of OH and O in reactions (6.24) and (6.25) are considered according to reactions (6.4), (6.5), and (6.6). The first case in which no interaction is considered leads to a very high coverage compared to the experimental values. If only the effect of OH interaction is included, the simulated coverage catches the experiment at low and medium potentials, but gives rise to higher values at high potentials. Adding the effect of O interaction improves the simulation results even further, and a very good agreement is obtained over the whole studied range of electrode potentials.

The reason for the improvement in the coverage predictions seems to be in the change of energetics on the reaction rates. At low potentials or at low coverages, the adsorbates are well spread on the surface such that they are at far distances from each other. Each elementary reaction proceeds practically on a local site and isolated from other species. As the coverage increases, elementary reactions are no longer isolated, but occur in the presence of other adsorbates, which can affect the rate of reaction.

6.8 Summary

In this chapter, a special computational approach was first developed to perform quantum calculations in a consistent way for locating transition states of ET reactions. Next, activation energies for six electron transfer reactions were computed, three elementary and three interaction reactions, as a function of the electrode potential. Interaction reactions were added to the previously developed chemical mechanism of PEM fuel cells because of their crucial role at high potentials.

As an application of the calculated kinetic rate, the results of Monte-Carlo simulations were presented. The coverage of O containing adsorbates obtained by these simulations for the water discharge mechanism was compared with the experimental measurements. It was shown that interaction effects can be quite significant.

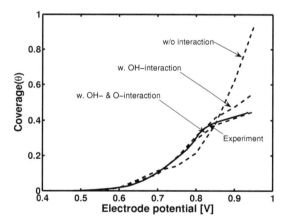

Figure 6.23: Comparison of the coverage of O containing species versus electrode potential for water discharge mechanism [101].

Chapter 7

Prefactors of ET Reactions

In this chapter, the pre-exponential factors are computed for the three elementary electron transfer reactions, which are present in the chemical mechanism of Chapter 3. First, a theoretical framework is developed that, based on the harmonic model, allows one to compute the pre-factors using quantum calculations. The harmonic model requires the mass matrix of the system to be estimated. A major difficulty that had to be overcome was to calculate mass matrices for molecular systems with dynamic constraints. The constraints in our systems are essentially those in the structure of the hydronium ion. The conventional way is to extract the mass matrix for such geometrically constrained systems from the expression of the kinetic energy, a process that becomes tedious and error prone for large systems.

Instead of following the conventional path, we extended a method for the computation of the mass matrix that avoids deriving the kinetic energy expression in a way that it can also consider systems that are subject to constraints [15]. Using this methodology, pre-exponential factors are estimated for three elementary reactions, assuming the same value for the forward and backward directions. In Appendix B, it is proved that these two values are equal within the Local Reaction Center theory, at least at standard and equilibrium conditions.

7.1 Theoretical Background

In the classical Marcus theory for ET processes, the potential energy surface is a quadratic function of the nuclear coordinates. That leads to a harmonic model for the system at the ground state [60]. For Marcus' model, the rate constant in the transition state theory can be written as [37]:

$$k_{\text{TST}} = A \, \exp\left(-\frac{E_{\text{act}}}{k_{\text{B}}T}\right) \tag{7.1}$$

in which E_{act}, k_{B}, and T are the activation energy, the Boltzmann constant, and the temperature.

Within the transition state theory, the pre-exponential factor can be estimated from the equilibrium distribution of states in phase space [92]. This gives an estimation of A based on the reorganization energies λ_μ associated with the normal modes μ:

$$A = \frac{1}{2\pi}\sqrt{\frac{\sum_\mu \omega_\mu^2 \lambda_\mu}{\lambda}} \tag{7.2}$$

where ω_μ is the vibrational frequency corresponding to mode μ. Equation (7.2) states that the overall frequency factor of reaction is the square root average of all modal frequencies weighted by their contribution to the reorganization energy λ. This provides a straightforward means for the calculation of pre-exponential factors within the assumptions of the transition state theory [92]. Traditionally, the reorganization energy is divided into two parts, one from the internal degrees of freedom of the reaction complex and the other from the normal modes associated with the solvent [60, 61, 72]. In our ab-initio computations, the solvent molecules are explicitly simulated. Hence, in the methodology that is developed here, no distinction will be made between the internal and solvent modes.

7.1.1 Modal Analysis and Pre-Exponential Factors

We seek a computational framework through which one can evaluate the pre-exponential factor of electron-transfer reactions in equation (7.1) using the approximation given by equation (7.2). To this end, two components have to be determined, the normal mode frequencies ω_μ and their corresponding reorganization energies λ_μ.

Assuming a harmonic model for the dynamics of the ET process, λ_μ is given by [70]

$$\lambda_\mu = \frac{1}{2}\bar{k}_\mu[(Q_\mu^\circ)_r - (Q_\mu^\circ)_p]^2 \tag{7.3}$$

$(Q_\mu^\circ)_r$ and $(Q_\mu^\circ)_p$ are the equilibrium values of normal mode μ corresponding to the ground states of reactants and products. $\bar{k}_\mu$ is an average force constant for the normal mode μ in terms of the force constants of the reactants k_μ^r and those of the products k_μ^p [70]:

$$\bar{k}_\mu = \frac{2k_\mu^r k_\mu^p}{k_\mu^r + k_\mu^p} \tag{7.4}$$

The total reorganization energy λ in equation (7.2) is the sum of the individual λ_μ's [70, 92]:

$$\lambda = \sum_\mu \lambda_\mu \tag{7.5}$$

Therefore, once the frequencies and force constants are known, the pre-exponential factor can be readily computed. This process requires certain computational steps explained below. To reach a harmonic description of the underlying dynamics, one has to determine the mass matrix $\mathbf{M}$ and the stiffness matrix $\mathbf{K}$ appearing in

$$\mathbf{M} * \ddot{\mathbf{X}} + \mathbf{K} * \mathbf{X} = 0 \tag{7.6}$$

where $\mathbf{X}$ is the vector of degrees of freedom in the structure of the reaction complex, and * represents matrix multiplication. Note that in general, $\mathbf{X}$ is different from the normal modes vector $\mathbf{Q}$. Once both matrices $\mathbf{M}$ and $\mathbf{K}$ are known, a suitable transformation can be applied to the system (7.6) to reach the normal mode representation of the system dynamics.

The quantum simulations of ET reactions always start with specified degrees of freedom. Thus, the vector $\mathbf{X}$ is a known quantity throughout the calculations. Also, the stiffness matrix $\mathbf{K}$ is directly accessible from the ground state computation of the studied system. In contrast, the mass matrix $\mathbf{M}$ is not a direct result of the calculations otherwise needed. Similar to the stiffness matrix $\mathbf{K}$, the mass matrix $\mathbf{M}$ depends on geometry as well as on the chemical composition of the precursor. One reasonable question here is if it would be possible to use the frequency analysis of the Gaussian software to assist in determining $\mathbf{M}$. It is worth knowing that the usual frequency analysis of Gaussian assumes a gas model, in which all the atoms within the molecule are allowed to move.

A modification to this approach is to fix certain atoms by specifying a proper keyword in the input file. This way, one can prevent, for example, the surface atoms from generating internal vibrational modes. Thereby, the corresponding phonon modes will be eliminated from the frequency analysis, and this reduces the dimension of the mass matrix in agreement with a constrained geometry. However, not every structural constraint in PDAE calculations corresponds to surface atoms. For the reaction complex for instance, not the atom positions, but certain bond lengths and angles have to be constrained. In other words, unlike the surface atoms, the atoms in the reaction complex cannot by frozen. Therefore, it is required to develop a general method capable of compiling the mass matrix associated with any arbitrary set of structural constraints, the geometry, and the chemical composition of the system.

The basic concepts for such a method are described by Wilson *et al.* [110]. Below, their approach to find the mass matrix is presented. Next, we discuss an extension to the theory that is capable of determining the mass matrix of a system subject to structural constraints.

7.1.2   Mass Matrix in Terms of Internal Coordinates

Internal Coordinates and Atomic Displacements. The material and terminology presented in this section is based on the work by Wilson *et al.* [110]. The internal coordinates and atomic displacements can be related to each other. A molecule with N number of atoms can have $n = 3N - 6$ internal degrees of freedom (DOF) at maximum. Note that internal DOF do not include the total displacement, rotation, and external vibration of the molecule. These $3N - 6$ degrees are enough to describe the internal geometry of the molecule, *i.e.* the position of all atoms with respect to each other. Internal vibrations of a molecule are essential in describing the internal energy of the molecule, and can be described using a system of internal coordinates. Since the kinetic energy is more conveniently determined in terms of Cartesian coordinates, a relationship between these two coordinate spaces is vital to establish a dynamic model. Due to the infinitesimal nature of vibrations, one can assume a linear mapping between internal coordinates $\mathbf{t} = \{t_i\}$ having variation S_{t_i} with an atom α undergoing displacement vector $\boldsymbol{\varrho}_\alpha$. Vector $\boldsymbol{\varrho}_\alpha \in \mathbb{R}^3$ contains Cartesian displacement coordinates for atom α. The relationship for the total change of coordinate t_i due to the possible displacement of all atoms is then given by [90, 110]

$$S_{t_i} = \sum_{\alpha=1}^{N} \boldsymbol{\varrho}_\alpha^T \mathbf{s}_{t_i \alpha} \qquad i = 1, ..., n \tag{7.7}$$

where the coupling displacement vector $\mathbf{s}_{t_i \alpha}$ is the direction of atom α that will create the largest increase in S_{t_i} when all the other atoms are fixed at their equilibrium positions. The magnitude of $\mathbf{s}_{t_i \alpha}$ equals the amount of increase in S_{t_i} caused by a unit displacement of atom α in the optimum direction. By specifying all vectors $\{\mathbf{s}_{t_i \alpha}\}$ for internal coordinate t_i, the mapping is completely determined.

One of the widely used internal coordinate systems is the Z-matrix description of the geometry of a molecule, in which there are three types of coordinates: bond lengths, bond angles, and dihedral angles.

Bond Length. Let S_t define the increase in the interatomic distance of two atoms 1 and 2 as shown in Fig. 7.1. Let $\mathbf{e}_{\alpha\beta}$ denote the unit vector from atom α to atom β, and $r_{\alpha\beta}$ the magnitude of the bond length. The maximum increase in S_t is obtained by moving each atom along the connecting line in a direction away from the other atom. Moreover, the magnitude of both $\mathbf{s}_{t1}$ and $\mathbf{s}_{t2}$ is unity since a unit displacement along the bond length will produce unit increase in S_t. All other $\mathbf{s}_{t\alpha}$, for which $\alpha \neq 1, 2$, do not contribute to any change in S_t and are hence equal to zero. Finally, if S_t is the stretching of the bond length between atom 1 and 2, one can write

$$\mathbf{s}_{t1} = \mathbf{e}_{21}$$
$$\mathbf{s}_{t2} = \mathbf{e}_{12} \tag{7.8}$$

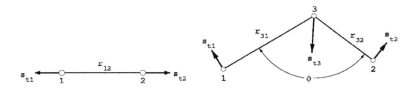

Figure 7.1: (left) Illustration of $\mathbf{s}_{t1}$ and $\mathbf{s}_{t2}$ as an increase in the bond length r_{12} between atoms 1 and 2, and (right) illustration of $\mathbf{s}_{t1}$, $\mathbf{s}_{t2}$, and $\mathbf{s}_{t3}$ as an increase in the bond angle ϕ formed between atoms 1, 2, and 3.

Bond Angle. In this case, S_t stands for the increase in the bond angle that is formed by three atoms 1, 2, and 3 (Fig. 7.1). The direction of $\mathbf{s}_{t1}$ is perpendicular to $\mathbf{e}_{31}$ and pointed outward to produce the maximum increase in the bond angle. Its magnitude is inversely proportional to r_{31} since a unit displacement of atom 1 along $\mathbf{s}_{t1}$ will cause ϕ to increase by $1/r_{31}$. $\mathbf{s}_{t2}$ can be obtained based on similar considerations. To find $\mathbf{s}_{t3}$, one notes that a rigid displacement of the whole structure leaves the bond angle unchanged. Assume that atom 3 is given a displacement along a certain direction $\mathbf{s}_{t3}$. By shifting the whole molecule along $-\mathbf{s}_{t3}$, atom 3 will be restored at its original

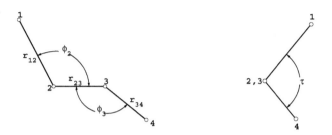

Figure 7.2: Illustration of dihedral angle τ between two planes formed by atoms 1, 2, 3 and 2, 3, 4.

position, while atoms 1 and 2 are displaced by $-\mathbf{s}_{t3}$. Hence, $\mathbf{s}_{t3}$ equals the negative sum of these vectors. Since $\mathbf{s}_{t1}$ and $\mathbf{s}_{t2}$ are already known, $\mathbf{s}_{t3}$ is found by

$$\mathbf{s}_{t3} = -\mathbf{s}_{t1} - \mathbf{s}_{t2} \tag{7.9}$$

Obviously, $\mathbf{s}_{t\alpha}$ for all other atoms, for which $\alpha \neq 1, 2, 3$, do not alter this bond angle and hence equal zero. It can be shown that [110]

$$
\begin{aligned}
\mathbf{s}_{t1} &= \frac{\cos\phi\,\mathbf{e}_{31} - \mathbf{e}_{32}}{r_{31}\sin\phi}, \\
\mathbf{s}_{t2} &= \frac{\cos\phi\,\mathbf{e}_{32} - \mathbf{e}_{31}}{r_{32}\sin\phi} \\
\mathbf{s}_{t3} &= \frac{(r_{31} - r_{32}\cos\phi)\mathbf{e}_{31} + (r_{32} - r_{31}\cos\phi)\mathbf{e}_{32}}{r_{31}r_{32}\sin\phi}
\end{aligned}
\tag{7.10}
$$

Dihedral Angle. This is the angle τ between two planes determined by atoms 1, 2, 3 and 2, 3, 4 as shown in Fig. 7.2. Following the methods discussed above, one can

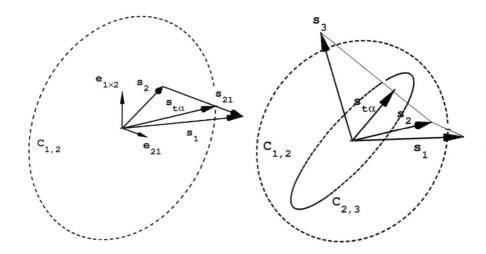

Figure 7.3: (left) Geometrical solution to a 2-fold constraint problem of combining s_1 and s_2, and (right) geometrical solution to a 3-fold constraint problem of combining s_1, s_2, and s_3.

find that [110]

$$s_{t1} = -\frac{e_{12} \times e_{23}}{r_{12} \sin^2 \phi_2}$$

$$s_{t2} = \frac{r_{23} - r_{12} \cos \phi_2}{r_{23} r_{12} \sin \phi_2} \frac{e_{12} \times e_{23}}{\sin \phi_2} + \frac{\cos \phi_3}{r_{23} \sin \phi_3} \frac{e_{43} \times e_{32}}{\sin \phi_3} \qquad (7.11)$$

$$s_{t3} = [(14)(23)]s_{t2}$$

$$s_{t4} = [(14)(23)]s_{t1}$$

The expressions (14) and (23) in the brackets in equation (7.11) stand for the permutation of the atom subscripts (1 and 4), and (2 and 3) in the relation defined for two vectors s_{t1} and s_{t2}.

G-Matrix and Kinetic Energy. The G-Matrix is defined as

$$G_{tt'} = \sum_{\alpha=1}^{N} \frac{1}{m_\alpha} \mathbf{s}_{t\alpha} \cdot \mathbf{s}_{t'\alpha} \tag{7.12}$$

where m_α is the mass of atom α. Note that by construction, matrix $\mathbf{G}$ is symmetric. Its components $G_{tt'}$ are related to the coupling displacement elements $\{\mathbf{s}_t\}$ between the internal coordinates t and t'. The significance of $\mathbf{G}$ becomes clear when the kinetic energy of internal vibrations is written explicitly as [110]

$$T = \frac{1}{2} \sum_{tt'} (G^{-1})_{tt'} \dot{S}_t \dot{S}_t \tag{7.13}$$

Equation (7.13) reveals that mass matrix $\mathbf{M}$ equals the inverse of matrix $\mathbf{G}$

$$\mathbf{M} = \mathbf{G}^{-1} \tag{7.14}$$

This outlines the mechanistic of computing $\mathbf{M}$:

1. Set up vectors $\{\mathbf{s}_{t\alpha}\}$ by analyzing the geometry definition of the system for each internal coordinate t and each atom α,

2. Compile matrix $\mathbf{G}$ using equation (7.12),

3. Invert $\mathbf{G}$ to obtain the mass matrix $\mathbf{M}$ that corresponds to the internal coordinates.

If an atom α participates in the definition of a certain coordinate t more than once, there will be more than one $\mathbf{s}_{t\alpha}$, namely $\mathbf{s}_{t\alpha}^{(1)}$ and $\mathbf{s}_{t\alpha}^{(2)}$ and so on. This series of vectors $\mathbf{s}_{t\alpha}^{(i)}$ with $i = 1..n \geq 2$ have to be combined to obtain the net $\mathbf{s}_{t\alpha}$. In this situation, the structure of the molecule can vary only in a way or ways to satisfy the applied constraints. Because neither Ref. [110] nor the literature provided the solution to this situation, we developed a methodology to obtain $\mathbf{s}_{t\alpha}$ in the presence of structural constraints [15].

7.2   Constraints in Mass Matrix Calculations

Case $n = 2$. First, we consider the case of $n = 2$ (2-fold constraint), in which an internal coordinate t and the atom α appear together in more than one line in the z-matrix definition of the molecule geometry. Let $\mathbf{s}_1$ and $\mathbf{s}_2$ denote two corresponding $\mathbf{s}_{t\alpha}$'s

$$
\begin{aligned}
\mathbf{s}_1 &= \mathbf{s}_{t\alpha}^{(1)} \\
\mathbf{s}_2 &= \mathbf{s}_{t\alpha}^{(2)}
\end{aligned}
\tag{7.15}
$$

We seek $\mathbf{s}_{t\alpha}$ by combining the information contained in $\mathbf{s}_1$ and $\mathbf{s}_2$. The direction of $\mathbf{s}_{t\alpha}$ is, by definition, a direction that maximizes S_t, and its magnitude is given by the change in S_t that corresponds to a unit displacement of atom α in the optimum direction. Obviously, the change in S_t should be the same for both definitions 1 and 2 if the constraint above is to be respected, *i.e.*

$$
\mathbf{s}_1 \cdot \mathbf{e}_x = \mathbf{s}_2 \cdot \mathbf{e}_x \; ,
\tag{7.16}
$$

where $\mathbf{e}_x$ is the unit vector along $\mathbf{s}_{t\alpha}$. We adopt a geometrical approach to solve this problem using Fig. 7.3. The constraint equation (7.16) implies that the projection of both $\mathbf{s}_1$ and $\mathbf{s}_2$ on $\mathbf{s}_{t\alpha}$ be the same, which leads to

$$
\mathbf{s}_{t\alpha} \cdot (\mathbf{s}_1 - \mathbf{s}_2) = 0
$$

or

$$
\mathbf{s}_{t\alpha} \cdot \mathbf{s}_{21} = 0 \; .
\tag{7.17}
$$

Equation (7.17) implies that $\mathbf{s}_{t\alpha}$ lies on a plane whose normal vector is given by $\mathbf{s}_{21}$ or its corresponding unit vector $\mathbf{e}_{21}$. One notes that the magnitude of S_t is maximized when $\mathbf{s}_{t\alpha}$ lies in the plane of $\mathbf{s}_1$ and $\mathbf{s}_2$ too. The optimum direction $\mathbf{e}_x$ is perpendicular to both the normal of the plane formed by $\mathbf{s}_1$ and $\mathbf{s}_2$, given by $\mathbf{e}_{1\times2}$, and also to $\mathbf{s}_{21}$

$$
\mathbf{e}_x = \mathbf{e}_{1\times2} \times \mathbf{e}_{21} \; .
\tag{7.18}
$$

Having determined the direction of $\mathbf{s}_{t\alpha}$, its magnitude is obtained by evaluating the projection of $\mathbf{s}_1$ or $\mathbf{s}_2$ on this direction using equation (7.16). Since two vectors $\mathbf{s}_1$ or $\mathbf{s}_2$ are always distinguishable, one can always find $\mathbf{s}_{t\alpha}$ for a 2-fold constraint. Physically, this situation means that an optimum direction for a 2-fold constraint can be generally found that maximizes the change in the internal coordinate t. It will be seen that this is not the case when $n > 2$, where the solution may be restricted or even impossible.

Case $n = 3$. In a 3-fold constraint, atom α appears three times in the z-matrix definition of the molecule geometry along with parameter t. For each constraint i, one finds the set of $\mathbf{s}_{t\alpha}^{(i)}, i = 1..3$, and the problem is to obtain $\mathbf{s}_{t\alpha}$ from these vectors. Once again, a graphical approach is exploited to find the solution (Fig. 7.3). The 3-fold constraint is given by

$$\mathbf{s}_1 \cdot \mathbf{e}_x = \mathbf{s}_2 \cdot \mathbf{e}_x = \mathbf{s}_3 \cdot \mathbf{e}_x . \tag{7.19}$$

One can use the solution method of $n = 2$ for $\mathbf{s}_1$ and $\mathbf{s}_2$, and for $\mathbf{s}_2$ and $\mathbf{s}_3$ separately to determine two planes $C_{1,2}$ and $C_{2,3}$, respectively. The final solution vector $\mathbf{s}_{t\alpha}$ must lie on both planes, or consequently on the intersection line between them. This way, the direction $\mathbf{e}_x$ is computed by

$$\mathbf{e}_x = \mathbf{e}_{21} \times \mathbf{e}_{32} . \tag{7.20}$$

The magnitude of $\mathbf{s}_{t\alpha}$ is determined by projecting any of $\mathbf{s}_{t\alpha}^{(i)}$ on $\mathbf{e}_x$ as before. Note that in general, the direction $\mathbf{e}_x$ cannot be optimized to maximize S_t as in the $n = 2$ case, rather it is fixed by equation (7.20). However, the optimization of $\mathbf{e}_x$ becomes possible if $\mathbf{e}_{21}$ and $\mathbf{e}_{32}$ are collinear. In this special case, the two planes $C_{1,2}$ and $C_{2,3}$ are the same, and the 3-fold constraint collapses to $n = 2$. Hence, for a 3-fold constraint, the answer is either a restricted or a full solution vector.

Cases $n > 3$. To search for a possible solution in these cases, one can find a candidate $\mathbf{e}_x^{(1)}$ for $\mathbf{e}_x$ by analyzing the first three vectors $\mathbf{s}_{t\alpha}^{(i)}$, $i = 1..3$. Then, by replacing the third vector $i = 3$ with another vector from set $i = 4..n$, an alternative $\mathbf{e}_x^{(2)}$ for $\mathbf{e}_x$ is obtained. Finally, a solution for $\mathbf{e}_x$ exists if all $\mathbf{e}_x^{(j)}$'s, determined in sequence, are collinear. Otherwise no solution exists. At this point, we can examine the physical meaning of the cases, in which there is no solution for $\mathbf{s}_{t\alpha}^{(i)}$. Assume there is only one degree of freedom for which one finds no solution for $\mathbf{e}_x$, and therefore leads to using a zero vector for $\mathbf{s}_{t\alpha}$. Equation (7.12) then returns a zero value for $G_{tt'}$ corresponding to this atom. A zero value in matrix $\mathbf{G}$ corresponds to an infinite value in matrix $\mathbf{M}$ according to equation (7.14). Physically, an infinite mass is a frozen object that cannot move. This is in line with the expectation that if atom α is over-constrained, it will have no freedom to vibrate.

7.3 Normal Mode Transformation

As mentioned earlier, the stiffness matrix $\mathbf{K}$ is directly determined from quantum simulations. The mass matrix $\mathbf{M}$ is calculated by the approach developed in the previous section. The next step is to transform the dynamic equation (7.6), which is in terms of the internal coordinates $\mathbf{X}$, to a dynamic equation in terms of the normal coordinates $\mathbf{Q}$, required by equations (7.2) to (7.5). The goal is to find a suitable transformation $\mathbf{U}$ that, once applied to equation (7.6), gives a set of decoupled equations in terms of independent modes $\mathbf{Q}$. This is a well-known problem in classical as well as in quantum mechanics (see for example Goldstein [45]). The transformation $\mathbf{U}$ is obtained by taking the following steps.

- Find transformation $\mathbf{V}_M$ that diagonalizes matrix $\mathbf{M}$ to $\mathbf{D}$

$$\mathbf{D} = \mathbf{V}_{\mathrm{M}}^{-1} * \mathbf{M} * \mathbf{V}_{\mathrm{M}} . \tag{7.21}$$

This step transforms equation (7.6) into

$$\mathbf{D} * \ddot{\mathbf{X}}_{\mathrm{M}} + \mathbf{K}_{\mathrm{M}} * \mathbf{X}_{\mathrm{M}} = 0 . \tag{7.22}$$

with $\mathbf{X}_{\mathrm{M}}$ and $\mathbf{K}_{\mathrm{M}}$ being the transformed $\mathbf{X}$ and $\mathbf{K}$, respectively.

- Scale $\mathbf{X}_M$ coordinates by the inverse of the diagonal elements of D_{ii} to make matrix $\mathbf{D}$ the identity matrix

$$I = \mathbf{V}_D^{-1} * \mathbf{D} * \mathbf{V}_D \ . \tag{7.23}$$

Consequently, the dynamic equation of the system reads

$$I * \ddot{\mathbf{X}}_D + \mathbf{K}_D * \mathbf{X}_D \ = \ 0$$

$$\text{or}$$

$$\ddot{\mathbf{X}}_D + \mathbf{K}_D * \mathbf{X}_D \ = \ 0 \ . \tag{7.24}$$

$\mathbf{X}_D$ and $\mathbf{K}_D$ are, accordingly, the representations of coordinates and stiffness matrix in this case.

- Determine transformation $\mathbf{V}_K$ that diagonalizes $\mathbf{K}_D$

$$\mathbf{K}_K = \mathbf{V}_K^{-1} * \mathbf{K}_D * \mathbf{V}_K \tag{7.25}$$

Note that this latter transformation leaves the decoupled acceleration term in equation (7.24) undisturbed, while it decouples the displacement term. The final dynamic equation thus becomes

$$\ddot{\mathbf{Q}} + \mathbf{K}_K * \mathbf{Q} \ = \ 0 \tag{7.26}$$

- Compute the desired transformation $\mathbf{U}$ as the combination of the three successive transformations described above.

$$\mathbf{Q} \ = \ (\mathbf{V}_M * \mathbf{V}_D * \mathbf{V}_K)^{-1} * \mathbf{X}$$

$$\mathbf{Q} \ = \ \mathbf{U} * \mathbf{X} \ . \tag{7.27}$$

Using transformation (7.27), one can transform the initial harmonic equation (7.6) into the normal mode equation (7.26), which is suitable for the calculations of pre-exponential factors from equation (7.2).

7.4 Estimation of Pre-Exponential Factors

In Chapter 6, the following three ET reactions were simulated, and their activation energies were computed.

$$OH_{(ads)} + H^+ + e^- \rightleftharpoons H_2O, \tag{6.1}$$

$$O_{(ads)} + H^+ + e^- \rightleftharpoons OH_{(ads)}, \tag{6.2}$$

$$O_{2(ads)} + H^+ + e^- \rightleftharpoons O_2H_{(ads)}. \tag{6.3}$$

Here, in light of the harmonic model, the pre-exponential factors are estimated. Denoting the stiffness matrix of the reactants by $\mathbf{K}_r$, and that of the products by $\mathbf{K}_p$, we define the averaged stiffness matrix $\mathbf{K}$ similar to equation (7.4) as

$$\mathbf{K} = 2(\mathbf{K}_r^{-1} + \mathbf{K}_p^{-1})^{-1}. \tag{7.28}$$

The averaged mass matrix is defined as

$$\mathbf{M} = \frac{\mathbf{M}_r + \mathbf{M}_p}{2}. \tag{7.29}$$

Equations (7.28) and (7.29) are then used to establish the dynamic equation of the system in terms of internal coordinates according to equation (7.6). The results of the normal mode analysis for the above electron-transfer reactions are given in Tables 7.3 through 7.2. The first two rows in each table, $\mathbf{X}^\circ_r$ and $\mathbf{X}^\circ_p$, are the equilibrium values of internal coordinates for the reactants and products, respectively. The rows, denoted by $\mathbf{Q}^\circ_r$ and $\mathbf{Q}^\circ_p$, are the normal mode vectors at the equilibrium states. The two rows f_μ and λ_μ represent the normal mode frequencies and reorganization energies. Using equation (7.2), the pre-exponential factors are calculated in Table 7.4. The frequency factor A for each reaction is of the order of 10^{13} $1/s$ in agreement with other theoretical estimations in the literature [37], where a value equal to $k_B T/\hbar$ is suggested for the pre-exponential factor of adiabatic ET reactions. Note that the calculated values in Table 7.4, are within the harmonic model for the vibrations, and also they do not include the concentration effects, *i.e.* the low concentration of H^+ and other surface species. This means that interactions and solvation effects change the reaction frequency by affecting the probability of finding solvated species within the vicinity of the surface. The task of taking into account the concentration effects can be accomplished by either a continuum or molecular dynamics approach or a combination of both.

Table 7.1: Normal mode analysis of Pt-PtOH$_2$[H$_2$O]$_3$ corresponding to reaction (6.1)

	$\mu = 1$	$\mu = 2$	$\mu = 3$	$\mu = 4$	$\mu = 5$	$\mu = 6$	$\mu = 7$	$\mu = 8$
$\mathbf{X}^\circ{}_r$	2.2047	0.97831	1.0293	1.5382	0.977	2.7598	0.96788	0.9673
$\mathbf{X}^\circ{}_p$	1.9691	0.98538	1.503	1.0443	1.0153	2.5606	0.97014	0.96946
f_μ	226.32	232.13	192.44	106.59	40.68	9.3442	13.5109	12.1587
λ_μ	1.083e-4	1.6138e-06	6.56e-3	3.92e-05	0.49781	0.09696	0.07873	0.15822

Table 7.2: Normal mode analysis of Pt-PtOH[H$_2$O]$_3$ corresponding to reaction (6.2)

	$\mu = 1$	$\mu = 2$	$\mu = 3$	$\mu = 4$	$\mu = 5$	$\mu = 6$	$\mu = 7$
$\mathbf{X}^\circ{}_r$	1.9067	1.0058	1.6626	0.97526	2.7755	0.96767	0.96726
$\mathbf{X}^\circ{}_p$	1.8069	1.5307	1.0398	1.0203	2.547	0.9702	0.96951
f_μ	213.34	229.99	233.70	41.34	21.638	9.105	15.4742
λ_μ	5.862e-3	7.217e-05	5.684e-08	0.80429	0.14217	0.05567	0.1769

Table 7.3: Normal mode analysis of PtO-PtOH[H$_2$O]$_3^\dagger$ corresponding to reaction (6.3)

	$\mu = 1$	$\mu = 2$	$\mu = 3$	$\mu = 4$	$\mu = 5$	$\mu = 6$	$\mu = 7$	$\mu = 8$
$\mathbf{X}^\circ{}_r$	1.9011	1.0558	1.5212	1.4412	0.9790	2.7351	0.9678	0.9673
$\mathbf{X}^\circ{}_p$	1.9065	1.5383	1.4642	1.0254	1.0194	2.5451	0.97014	0.9694
$f_\mu^\ddagger$	232.14	226.40	189.66	43.8411	7.986	10.390	16.12	14.167
$\lambda_{\mu\ddagger}$	3.20e-07	1.122e-04	0.010678	0.61409	0.017632	5.718e-03	2.343e-03	0.12688

† For this system, two zero frequencies were eliminated from the vibrational analysis. This reduced the number of degrees of freedom from 10 to 8.
‡ Normal mode frequencies are in units of THz = 10^{12} 1/sec, and reorganization energies are in eV.

Table 7.4: Pre-exponential factors of electron transfer reactions (6.1)-(6.3)

Reaction	A [1/s]
$OH_{(ads)} + H^+ + e^- \rightleftharpoons H_2O$	36.536×10^{12}
$O_{(ads)} + H^+ + e^- \rightleftharpoons OH_{(ads)}$	38.522×10^{12}
$O_{2(ads)} + H^+ + e^- \rightleftharpoons O_2H_{(ads)}$	45.336×10^{12}

Appendix A

Extensions to LRC Theory

In this appendix, we try to extend the local reaction center theory (LRC) such that it can be applied to systems that have more than a few, and ideally an arbitrary number of catalyst atoms. This extension is required if systems with more realistic representations of the catalyst surface have to be studied. In chapter 5, it was shown by direct quantum computations that the properties of a molecule quickly become dominated by the bulk properties of large catalyst atoms such as platinum.

Large numbers of Pt atoms force the electrochemical activity of a molecule, that is characterized by its ionization potential (IP) and electron affinity (EA), to approach the work function of bulk platinum. This is an important issue in the LRC theory because IP and EA are compared to the work function of the electrode to check if the electron transfer condition is satisfied. Modifications to the LRC theory and/or its computational implementation are thus necessary before it can be applied to systems having many catalyst atoms. However, the validity and range of applications of any suggested improvement should be carefully examined by using electron transfer theories and computational tools. Here, we investigate two approaches: i) using point-charges to computationally excite the catalyst Pt cluster, and ii) using a different reaction center when calculating the IP or EA to check the electron transfer condition. Although at present none of these two ideas can be considered as a full improvement to the LRC method, they can render the core concepts for future developments. Moreover, the quantum response of the systems studied here provide valuable insights into the computational aspects of electrochemistry.

A.1 Exciting Electrode by External Charges

Catalysts and electrodes are mostly made of metals, or at least covered by metal atoms. The metallic bond is characterized by its delocalized electronic charge, which leads to high conductivity. High conductivities of metals are the main rationale for many basic principles in classical electrostatics, particularly, the vanishing of electric field inside bulk metals. Any amount of extra charge in metals then has to distribute on the surface, and does not occupy any inside location. The surface charge manifests itself as the electrostatic potential of metals. Therefore, using point charges on the surface of an electrode in the quantum computations is in a way consistent with the physical origins of the electrode potential. Additionally, surface point charges in computations build up an electric field that, along with its interaction with the adsorbates, can ideally represent the double layer on electrified surfaces.

In this section, we quantify the effect of surface charges on the IP of a few molecules, which contain different numbers of Pt atoms. The IP is calculated within the assumptions of the LRC theory after the ground state structure of the molecule is determined. This implies that the surface charges affect the IP not only directly through modifying the charge distribution, but also indirectly through changing the ground state structure. The adsorbate is an $OH[H_2O]_3$ molecule which undergoes the following oxidation step

$$Pt_n OH[H_2O]_3 \rightarrow Pt_n O + H[H_2O]_3^+ + e^- \tag{A.1}$$

Here Pt_n stands for the metal cluster used in the quantum simulations. These clusters are denoted by $n = 1$, $(3, 1)$, $(7, 3)$, where, for example, $Pt_{3,1}$ is a cluster having 3 atoms in the first layer close to the adsorbates, and 1 in the second layer. The point charge for Pt_1 cluster was placed 0.7 Å below the Pt atom, because no electrode surface could be defined properly for this case (Figures A.1). For the other cases, the surface charges were placed midway between any two surface atoms (Figures A.2 and A.3), according to the model of exciting a metal cluster by surface charges that was explained earlier in this section.

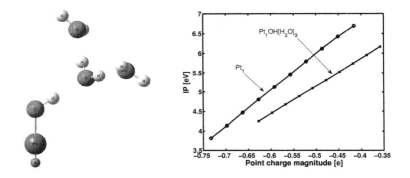

Figure A.1: Exciting Pt cluster with a point charge in $Pt_1OH[H_2O]_3$, (left) molecular structure and the point charge, denoted by X, (right) IP of Pt_1 and $Pt_1OH[H_2O]_3$ as a function of the charge magnitude.

Computations were performed at the B3LYP level of theory using 6-31G*/LANL2MB basis sets for O,H/Pt atoms. For each n, two molecules were considered: a bare Pt cluster, and a Pt cluster with an adsorbate molecule $OH[H_2O]_3$. The Pt-Pt bond lengths were fixed at the bulk value 2.775 Å. The adsorbate coordinates, on the other hand, were allowed to vary in determining the structure of the ground state. In each case, after the optimization step, the IP was calculated as the change in the electronic energy between the neutral and the positive ion. The magnitude of the point charges were chosen such that the IP values cover the full range of possible values for the work function of the cathode electrode in a fuel cell, say about 4.0-6.5 eV in absolute scale.

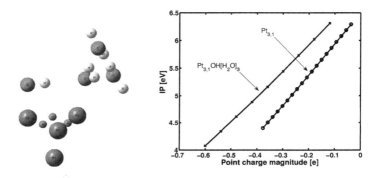

Figure A.2: Exciting Pt cluster with point charges in $Pt_{3,1}OH[H_2O]_3$, (left) molecular structure and the point charges, denoted by X, (right) IP of $Pt_{3,1}$ and $Pt_{3,1}OH[H_2O]_3$ as a function of the total charge magnitude.

For the Pt_1 cluster, the IP of the complete molecule is smaller than that of the bare cluster. The reverse trend is seen for $Pt_{3,1}$ and $Pt_{7,3}$ clusters. This difference in behavior can be attributed to the special position of the point charge in the Pt_1 cluster. The difference in IP between the full molecule and the Pt cluster with varying point charge is non-uniform. In the the largest system, $Pt_{7,3}$, the adsorbate only causes a constant vertical change in the IP over the whole range of point charges. If a very large Pt cluster is used, one can expect a quite predictable and uniform change in the IP due to the presence of the adsorbates.

The most important observation is that in all cases, the IP can be adjusted to any value in the potential range of interest through using relatively small amounts of surface charges, which in sum are less than ± 1 e. This means that unlike in the LRC theory, the ground state IP of a molecule is no longer a direct outcome of the quantum computations, but rather an input parameter chosen by the user. Any choice for surface charges has then to be justified to give compatible results when using different

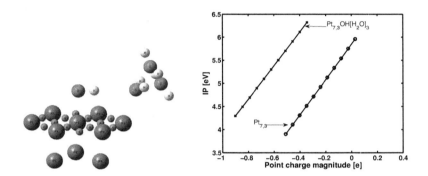

Figure A.3: Exciting Pt cluster with point charges in $Pt_{7,3}OH[H_2O]_3$, (left) molecular structure and the point charges, denoted by X, (right) IP of $Pt_{7,3}$ and $Pt_{7,3}OH[H_2O]_3$ as a function of the total charge magnitude.

Pt clusters. In conclusion, perturbing the initial system through introducing point charges leads to a new degree of freedom in computations, and requires to be explored further, especially from a theoretical point of view.

A.2   Isolating Reaction Center in ψ Calculations

Another possible extension emerges by focusing on the formulation of the LRC theory as presented in Chapter 4. According to that picture, determining transition states involves minimizing the activation energy φ, while at the same time satisfying the ET condition on the chemical potential ψ. From a computational point of view, catalyst atoms are needed in the system mainly to lower φ, but can be omitted in the computation of ψ, which should preferably represent the IP or EA of the reaction center (RC). If IP or EA are contaminated by the presence of too many catalyst atoms, an alternative approach might be to calculate ψ for the adsorbates without considering the catalyst atoms. This calculation of an isolated molecule is essentially valid for an outer-sphere electron-transfer reaction. Here, we study the application of the same approach to inner sphere reactions in the framework of the LRC theory.

Let us denote φ by φ_{sys}, and ψ by ψ_{RC} to emphasize the domains for which they have been computed. The Lagrange function (4.4) is generalized to equation

$$\mathcal{L}(\mathbf{x}, \mu) = \varphi_{\text{sys}}(\mathbf{x}) - \mu[\psi_{\text{RC}}(\mathbf{x}) - eE_e] \tag{A.2}$$

which can be solved by the CICV search method of Chapter 4.

Here we present and discuss the results of applying formulation (A.2) to $\text{Pt}_n\text{OH}[\text{H}_2\text{O}]_3$ with $n = 2$, $(3, 1)$, $(7, 3)$ to simulate reaction (A.1). In the sense of formulation (A.2), we chose for all cases

$$\begin{aligned} \text{sys} &= \text{Pt}_n\text{OH}[\text{H}_2\text{O}]_3 \\ \text{RC} &= \text{OH}[\text{H}_2\text{O}]_3 \end{aligned} \tag{A.3}$$

The ground state IP of the $\text{OH}[\text{H}_2\text{O}]_3$ molecule using different basis sets is given in Table A.1. Although the IP values are close, they are not exactly the same but differ

Table A.1: Ground state IP of $\text{OH}[\text{H}_2\text{O}]_3$ determined by B3LYP calculations using different basis sets.

Pt cluster	Basis set .	IP [eV]
$\text{Pt}_{3,1}$	6-31G*/LANL2MB	11.60984187
$\text{Pt}_{3,1}$	6-311G**/LANL2DZ	11.81741000
$\text{Pt}_{7,3}$	6-31G*/LANL2MB	11.57082574

by a few tenths of electron volts. Interestingly, using the same basis sets, the IP of an isolated $\text{OH}[\text{H}_2\text{O}]_3$ on $\text{Pt}_{3,1}$ is approximately 0.04 eV higher than on $\text{Pt}_{7,3}$. This difference is caused by the structure of the ground state being slightly different on those Pt clusters. The calculated IP using basis sets 6-311G**/LANL2DZ is about 11.82 eV, which is 0.19 eV higher than the IP determined by using 6-31G*/LANL2MB basis sets. This demonstrates the impact of basis sets on the calculations of ionization potentials (see Chapter 5). Next, the ψ-dependent activation energy (ψDAE) curves for

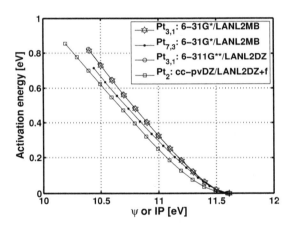

Figure A.4: Activation energy for oxidation of $OH[H_2O]_3$ molecule as a function of its IP using different Pt clusters and basis sets. All IP's have been shifted to 11.61 eV, which is the ground state IP of $OH[H_2O]_3$: B3LYP/6-31G*.

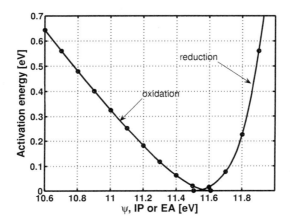

Figure A.5: Activation energy for reaction $OH_{(ads)} \rightleftharpoons O_{(ads)} + H^+_{(aq)} + e^-$ using $Pt_{3,1}OH[H_2O]_3$: B3LYP/6-31G*/LANL2MB.

oxidation reaction (A.1) were calculated by the CICV method. Figure A.4 illustrates the ψDAE curves, where all the curves have been adjusted to start from 11.61 eV on the horizontal axis (see Table A.1). The ψDAE curve for the same reaction using a Pt_2 cluster is also given for comparison purposes. For this last system, the reaction center was the whole molecule $Pt_2OH[H_2O]_3$, in accordance with the original implementation of the LRC theory.

Aside from the horizontal shift, the activation curves are in good agreement with each other, from which one can infer that isolating the reaction center from the catalyst atoms in IP calculations allows more uniform results to be obtained. For instance, the curves for $Pt_{3,1}OH[H_2O]_3$ molecule using two basis sets 6-31G*/LANL2MB and 6-311G**/LANL2DZ match perfectly. These promising results encouraged performing activation energy calculations for the reverse of reaction (A.1).

The ψDAE curve for the reduction step was calculated using $Pt_{3,1}OH[H_2O]_3$: B3LYP/6-31G*/LANL2MB, and is shown in Figure A.5 along with the oxidation curve. The ground state values of ψ are approximately 11.5 and 11.6 eV, which means both the curves start almost from the same ψ. Whether meaningful or not, this outcome should not be surprising, because the structure of the isolated $OH[H_2O]_3$ is not very different when it is taken from the optimized structures of $Pt_{3,1}OH[H_2O]_3$ and $Pt_{3,1}OH[H_2O]_3^+$ molecules. Similarity in the ground state structures has been the reason for the $OH[H_2O]_3$ molecule to have an IP very close to the EA of the $OH[H_2O]_3^+$ molecule. Although absent from the ψ calculations, a large electrode surface has strongly influenced the structure of the reaction center such that the ground state of ψ is almost indifferent to the ionic charge of the system. At the current stage, further investigations are necessary to improve the method of isolating reaction center and/or the interpretation of its results in order to transform it into a more satisfactory approach.

Finally, this question may arise how the activation energies would change if one, instead of removing all Pt atoms from ψ calculations, allowed a few of them to be present in the reaction center. We address this question below by presenting the activation curves for the oxidation reaction (A.1). The calculations were performed for a $Pt_{3,1}OH[H_2O]_3$ molecule, and four cases corresponding to different numbers of Pt atoms in the reaction center, that is

$$
\begin{aligned}
case\ 1 &: \mathrm{RC} = Pt_0OH[H_2O]_3 \\
case\ 2 &: \mathrm{RC} = Pt_1OH[H_2O]_3 \\
case\ 3 &: \mathrm{RC} = Pt_3OH[H_2O]_3 \\
case\ 4 &: \mathrm{RC} = Pt_{3,1}OH[H_2O]_3
\end{aligned}
\tag{A.4}
$$

The activation curves for the four cases in (A.4) are shown in Figure A.6, where the IP's have been adjusted such that all the curves start from 11.61 eV on the horizontal axis. The original ground state IP's are also shown in the figure. As the number of Pt atoms in the RC increases, the activation energies also increase, but

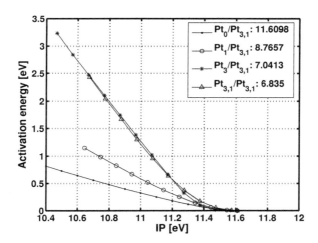

Figure A.6: Activation energy for oxidation of $OH[H_2O]_3$ molecule on a $Pt_{3,1}$ cluster as a function of its IP using different number of Pt atoms in the IP calculations. All IP's have been shifted to 11.61 eV, which is the ground state IP of $OH[H_2O]_3$: B3LYP/6-31G*.

reach a limiting case: cases 3 and 4 have clearly the same energetics at almost all values of IP. We now discuss the reason why the curves become steeper as the number of Pt atoms in the reaction center increases. Recall the dominating effect of large Pt atoms in the calculations of IP: more Pt atoms means less sensitivity to the removal of electrons from the whole system. The more Pt atoms are used in the RC, the less sensitive it will become to being ionized. Starting from any initial IP, the decrease in sensitivity demands the structure of the adsorbates to vary more and more to satisfy the ET condition, because the Pt atoms have fixed positions in the molecule. Larger variations in the structure from that of the ground state leads naturally to higher activation energies as depicted in Figure A.6. The question of exactly how many Pt atoms in a certain reaction center might lead to the most accurate activation energies has to be addressed through future studies.

Appendix B

Equality of ET Prefactors

In this appendix, it is shown that the pre-exponential factors of electron transfer reactions for the oxidation and reduction directions are equal under certain conditions. The equality specifically applies to the equilibrium and standard conditions, and it is subject to the assumptions of the Local Reaction Center theory.

B.1 Kinetics of Electron Transfer Reactions

Consider the general form of an electrochemical reaction, in which n electrons are exchanged between a reaction center (O and R) and an electrode with the corresponding forward and backward rate constants k_{red} and k_{oxi}

$$O + ne^- \overset{k_{red}}{\underset{k_{oxi}}{\rightleftharpoons}} R \qquad (B.1)$$

Here, we express the forward and backward rate constants, according to the transition state theory, in terms of the pre-exponentials A and the activation energies E_{act} [18]

$$k_{red} = A_{red} \exp\left(-\frac{E_{act}^{red}}{k_B T}\right) \qquad (B.2)$$

and

$$k_{\text{oxi}} = A_{\text{oxi}} \exp\left(-\frac{E_{\text{act}}^{\text{oxi}}}{k_{\text{B}}T}\right) \tag{B.3}$$

The *net cathodic rate* of reaction (B.1) is obtained by subtracting the forward and backward rates taking into account the concentration of the oxidized C_{red} and the reduced C_{oxi} centers. One can write

$$net\ cathodic\ rate = k_{\text{red}}C_{\text{red}} - k_{\text{oxi}}C_{\text{oxi}} \tag{B.4}$$

Equation (B.4) assumes that the concentration of electrons in the metal electrode is unity.

B.2   Equilibrium and Standard Conditions

By definition, the *equilibrium* condition is the state at which the forward and backward fluxes in equation (B.1) cancel each other [18]. Using equation (B.4), the equilibrium condition is obtained as

$$k_{\text{red}}C_{\text{red}} = k_{\text{oxi}}C_{\text{oxi}} \qquad @equilibrium \tag{B.5}$$

Now, if the equilibrium occurs when the concentrations are equal, i.e. when $C_{\text{red}} = C_{\text{oxi}}$, the rate constants are also equal.

$$k_{\text{red}} = k_{\text{oxi}} = k_0 \qquad @standard\ conditions \tag{B.6}$$

This situation is referred to as the *standard* conditions.

One implicit assumption in the derivations so far is that the rate constants are independent of the concentrations. However, the quantum computations are performed with a finite system, which already includes the information about coverage (concentrations) and either the presence or absence of interactions. Here, we keep the assumption of rate constants being independent of the concentrations [18].

The equilibrium electrode potential at the standard conditions is denoted by U_0. Below, the subscript 0 will be used for quantities which refer to the standard conditions.

B.3 Potential-Dependent Activation Energies

The activation energy E_{act} of an electron transfer reaction is the increase in the Gibbs free energy of the system at the transition state, denoted by #, with respect to that of the ground state

$$E_{act} = \Delta G^{\#} = G^{\#} - G^{\text{ground state}} \tag{B.7}$$

Here, we wish to express the activation energies at an arbitrary electrode potential U. In general, instead of U, the overvoltage $\eta = U - U_0$ is used in electrochemistry. Using η as the independent variable, and the Gibbs free energy at the standard potential as a constant, we can write the energy of the system at the transition state $G^{\#}$ as

$$G^{\#} = G_0^{\#} + g(\eta) \tag{B.8}$$

where $g(\eta)$ is an unknown function of the overpotential. Obviously, $g(\eta)$ has to vanish at the standard potential, that is

$$g(\eta)\big|_{\eta=0} = 0 \tag{B.9}$$

Substitution of equation (B.8) into equation (B.7) yields

$$\Delta G^{\#} = \Delta G_0^{\#} + g(\eta) \tag{B.10}$$

where $\Delta G_0^{\#}$ is the activation energy at the standard potential

$$\Delta G_0^{\#} = G_0^{\#} - G^{\text{ground state}} \tag{B.11}$$

Equation (B.10) shows that in order to determine the activation energy at an arbitrary potential U, which corresponds to an overpotential η, one only needs to determine the function $g(\eta)$. From a mathematical point of view, $g(\eta)$ all the terms in the Taylor expansions of $\Delta G^{\#}$, except the constant term. In the Butler-Volmer approximation, $g(\eta)$ is expressed by simple linear forms for oxidation and reduction steps

$$oxidation: \ g_{\text{oxi}}(\eta) = \alpha F \eta \tag{B.12}$$

and

$$reduction: \ g_{\text{red}}(\eta) = -\beta F \eta \tag{B.13}$$

where F is Faraday's constant, and α and β are two other constants, called the symmetry factors or the transition coefficients (see Chapter 3). These two factors determine the position of the transition state with respect to the ground states. α and β are not always independent constants, but under the assumptions that were discussed in Chapter 3, are related to each other through the following equation

$$\alpha + \beta = 1 \tag{B.14}$$

B.4 Rate Constants and Equilibrium Current

Consider the oxidation reaction rate constant given by equation (B.3), for which the activation energy can be obtained from equation (B.10)

$$\Delta G_{\text{oxi}}^{\#} = \Delta G_{0,\text{oxi}}^{\#} + g_{\text{oxi}}(\eta) \tag{B.15}$$

Using equation (B.15), the rate constant k_oxi takes the following form

$$
\begin{aligned}
k_\text{oxi} &= A_\text{oxi} \; \exp\left(-\frac{\Delta G^{\#}_{0,\text{oxi}}}{k_\text{B}T}\right) \exp\left(-\frac{g_\text{oxi}(\eta)}{k_\text{B}T}\right) \\
&= k_0 \quad \exp\left(-\frac{g_\text{oxi}(\eta)}{k_\text{B}T}\right)
\end{aligned} \tag{B.16}
$$

where k_0 is the rate constant at the standard and equilibrium conditions presented in equation (B.6). Similarly, the rate constant for the reduction step equals

$$
k_\text{red} = k_0 \; \exp\left(-\frac{g_\text{oxi}(\eta)}{k_\text{B}T}\right) \tag{B.17}
$$

The net cathodic current for an electrode with an active area A_e is the difference between the currents flowing through the electrode by the reduction and oxidation steps

$$
\begin{aligned}
i &= i_\text{red} - i_\text{oxi} \\
&= n \, F \, A_e \, (k_\text{red}C_\text{O} - k_\text{oxi}C_\text{R}) \\
&= n \, F \, A_e \, k_0 \left[C_\text{O} \exp\left(-\frac{g_\text{red}(\eta)}{k_\text{B}T}\right) - C_\text{R} \exp\left(-\frac{g_\text{oxi}(\eta)}{k_\text{B}T}\right) \right]
\end{aligned} \tag{B.18}
$$

B.5 Nernst Equation

In this section, we determine the equilibrium potential E_eq at any non-standard concentration of reactants and products. The equilibrium condition implies that the net current in equation (B.18) be zero

$$
i = 0 \implies k_\text{red}C_\text{O} = k_\text{oxi}C_\text{R} \tag{B.19}
$$

which can be simplified as

$$
g_\text{red}(\eta) - g_\text{oxi}(\eta) = k_\text{B}T \, \ln\left(\frac{C_\text{O}}{C_\text{R}}\right) \tag{B.20}
$$

Equation (B.20) expresses the difference in the activation energies of oxidation/reduction steps due non-standard concentrations of C_{red} and C_{oxi}. It is noteworthy that this result is independent of any electron transfer theory within which the $g_{red}(\eta)$ and $g_{oxi}(\eta)$ functions are obtained.

Here, we show that if the transition states are determined by any computational method that is based on the Local Reaction Center (LRC) theory, equation (B.20) will lead to the Nernst equation. According to the LRC theory, the ionization potential (IP) or the electron affinity (EA) of the reaction center at the transition state equals the work function of the electrode. For the n-electron electrochemical reaction (B.1) that happens at an electrode with the equilibrium potential U_{eq}, the electron transfer condition becomes:

$$\psi = n \ F \ U_{eq} \tag{B.21}$$

in which ψ denotes IP and EA of the system. In the LRC theory, ψ is determined from the energy of the reactants and that of the products at any point on the potential energy surface (PES)

$$\psi = G_{red}^{\#} - G_{oxi}^{\#} \tag{B.22}$$

At the transition state, equation (B.21) and equation (B.22) can be combined to give

$$G_{red}^{\#} - G_{oxi}^{\#} = n \ F \ U_{eq} \tag{B.23}$$

Equation (B.23) is valid at any equilibrium potential U_{eq} and hence at the standard potential U_0 too

$$G_{0,red}^{\#} - G_{0,oxi}^{\#} = n \ F \ U_0 \tag{B.24}$$

Now let's express the left hand side of equation (B.20) using equation (B.23), equation (B.24) and the definition of $g(\eta)$ in equation (B.8)

$$
\begin{aligned}
g_{\text{red}}(\eta) - g_{\text{oxi}}(\eta) &= (G_{\text{red}}^{\#} - G_{0,\text{red}}^{\#}) - (G_{\text{oxi}}^{\#} - G_{0,\text{oxi}}^{\#}) \\
&= (G_{\text{red}}^{\#} - G_{\text{oxi}}^{\#}) - (G_{0,\text{red}}^{\#} - G_{0,\text{oxi}}^{\#}) \\
&= n\ F\ U_{\text{eq}} - n\ F\ U_0 \\
&= n\ F\ (U_{\text{eq}} - U_0)
\end{aligned}
\tag{B.25}
$$

Equation (B.25) can be combined with equation (B.20) to arrive at the Nernst equation:

$$
U_{\text{eq}} = U_0 + \frac{k_{\text{B}}T}{n\ F} \ln\left(\frac{C_{\text{O}}}{C_{\text{R}}}\right)
\tag{B.26}
$$

It is seen that the Nernst equation is valid within the LRC theory too.

B.6 Thermodynamics of Electrochemical Reactions

In this section, we focus on the thermodynamics of the electrochemical reaction (B.1). The goal is to relate the ground state energy of the reactants and that of the products to the electrode potential U_{eq} at equilibrium condition. Using the expressions of the internal energy U and Gibbs free energy G, one can write

$$
dU = TdS - dW_{\text{electrical}} - PdV
\tag{B.27}
$$

and thus

$$
\begin{aligned}
dG &= dU - d(TS) + d(PV) \\
&= -SdT + VdP - dW_{\text{electrical}}
\end{aligned}
\tag{B.28}
$$

For constant temperature T and pressure P, equation (B.28) is simplified to

$$
dG = -dW_{\text{electrical}}
\tag{B.29}
$$

Equation (B.29) can be expressed in terms of the reactants' and products' energy (see Reference [77]).

$$\Delta G_{\text{rxn}} = -W_{\text{electrical}} \tag{B.30}$$

where ΔG_{rxn} is the total change in the Gibbs free energy of the system undergoing the electrochemical reaction (B.1). The electrical work in exchanging a charge of magnitude Q with an electrode with potential U_{eq} is

$$W_{\text{electrical}} = Q \; U_{\text{eq}} \tag{B.31}$$

When Q is carried only by electrons in an electrochemical reaction, it is equal to

$$Q = n \; F \tag{B.32}$$

Combining equation (B.30), equation (B.31) and equation (B.32) results in [77]

$$\Delta G_{\text{rxn}} = -n \; F \; U_{\text{eq}} \tag{B.33}$$

As mentioned above, ΔG_{rxn} is the change of the energy of the system, hence for reaction (B.1)

$$\Delta G_{\text{rxn}} = G_{\text{oxi}}^{\text{ground state}} - G_{\text{red}}^{\text{ground state}} \tag{B.34}$$

Finally, using equation (B.34) and equation (B.33) yields

$$G_{\text{red}}^{\text{ground state}} - G_{\text{oxi}}^{\text{ground state}} = n \; F \; U_{\text{eq}} \tag{B.35}$$

and for the standard conditions

$$G_{0,\text{red}}^{\text{ground state}} - G_{0,\text{oxi}}^{\text{ground state}} = n \; F \; U_0 \tag{B.36}$$

B.7 Rates at Standard & Equilibrium Conditions

In this section, the previous parts are combined to obtain important results about the rate constants at the equilibrium and standard conditions. To summarize, the main results are:

1. equilibrium condition, equation (B.5)

2. standard condition, equation (B.6)

3. definition of activation energy, equation (B.7)

4. electrode potential, defined in the LRC theory, as applied to U_0, equation (B.24)

5. thermodynamics of electrochemical reactions, equation (B.35)

We substitute all the information into the standard condition (B.6):

$$k_{\text{red}} = k_{\text{oxi}} \ @ \ standard \ \& \ equilibrium \ conditions \tag{B.37}$$

$$A_{\text{red}} \ \exp\left(-\frac{\Delta G^{\#}_{0,\text{red}}}{k_{\text{B}}T}\right) = A_{\text{oxi}} \ \exp\left(-\frac{\Delta G^{\#}_{0,\text{oxi}}}{k_{\text{B}}T}\right) \tag{B.38}$$

$$\Rightarrow \frac{A_{\text{red}}}{A_{\text{oxi}}} = \exp\left(\frac{\Delta G^{\#}_{0,\text{red}} - \Delta G^{\#}_{0,\text{oxi}}}{k_{\text{B}}T}\right)$$

$$= \exp\left(\frac{(G^{\#}_{0,\text{red}} - G^{\text{ground state}}_{0,\text{red}}) - (G^{\#}_{\text{oxi}} - G^{\text{ground state}}_{0,\text{oxi}})}{k_{\text{B}}T}\right)$$

$$= \exp\left(\frac{(G^{\#}_{0,\text{red}} - G^{\#}_{0,\text{oxi}}) - (G^{\text{ground state}}_{0,\text{red}} - G^{\text{ground state}}_{0,\text{oxi}})}{k_{\text{B}}T}\right) \tag{B.39}$$

$$= \exp\left(\frac{(n \ F \ U_0) - (n \ F \ U_0)}{k_{\text{B}}T}\right)$$

$$= 1$$

$$\therefore A_{\text{red}} = A_{\text{oxi}} \ @ \ standard \ \& \ equilibrium \ conditions \tag{B.40}$$

Here, the equality of pre-exponential factors was obtained because in the equilibrium condition (B.38), the change in the ground state energies was exactly canceled out by the change in the activated state energies through equation (B.24), which is used in the LRC theory.

Bibliography

[1] R. Adzic. In J. Lipkowski and P. N. Ross, editors, *Electrocatalysis*, page 197. Wiley-VCH: New York, 1998.

[2] A. V. Anantaraman and C. L. Gardner. Studies on ion-exchange membranes. part 1. effect of humidity on the conductivity of nafion(r). *Journal of Electroanalytical Chemistry*, 414:115–120, 1996.

[3] A.B. Anderson. Reactions and structures of water on clean and oxygen covered Pt(111) and Fe(100). *Surface Science*, 105(1):159–176, April 1981.

[4] A.B. Anderson. O_2 reduction and CO oxidation at the pt-electrolyte interface: The role of H_2O and OH adsorption bond strengths. *Electrochimica Acta*, 47(22-23):3759–3763, 2002.

[5] AB Anderson. Theory at the electrochemical interface: Reversible potentials and potential-dependent activation energies. *Electrochimica Acta*, 48(25-26):3743–3749, 2003.

[6] A.B. Anderson and T.V. Albu. Ab initio approach to calculating activation energies as functions of electrode potential - Trial application to four-electron reduction of oxygen. *Electrochemical Communications*, 1(6):203–206, 1999.

[7] A.B. Anderson and T.V. Albu. Catalytic effect of platinum on oxygen reduction: An ab initio model including electrode potential dependence. *Journal of Electrochemical Society*, 147(11):4229–4238, 2000.

[8] A.B. Anderson, Y. Cai, R.A. Sidik, and D.B. Kang. Advancements in the local reaction center electron transfer theory and the transition state structure in the first step of oxygen reduction over platinum. *Journal of Electroanalytical Chemistry*, 580(1):17–22, 2005.

[9] A.B. Anderson and D.B. Kang. Quantum chemical approach to redox reactions including potential dependence: Application to a model for hydrogen evolution from diamond. *Journal of Physical Chemistry A*, 102(29):5993–5996, 1998.

[10] A.B. Anderson and N.M. Neshev. Mechanism for the electro-oxidation of carbon monoxide on platinum, including electrode potential dependence: Theoretical determination. *Journal of Electrochemical Society*, 149(10):E383–E388, 2002.

[11] A.B. Anderson, N.M. Neshev, R.A. Sidik, and P. Shiller. Mechanism for the electrooxidation of water to OH and O bonded to platinum: Quantum chemical theory. *Electrochimica Acta*, 47(18):2999–3008, Jul 2002.

[12] AB Anderson, RA Sidik, J Narayanasamy, and P Shiller. Theoretical calculation of activation energies for $Pt + H^{+}(aq) + e^{-}(U) \leftrightarrow Pt - H$: Activation energy-based symmetry factors in the marcus normal and inverted regions. *Journal of Physical Chemistry B*, 107(19):4618–4623, 2003.

[13] A.B. Anderson, R.A. Sidik, J. Narayanasamy, and P. Shiller. Theoretical calculation of activation energies for Pt+H+(aq)+e(-)(U)¡-¿ Pt-H: Activation energy-based symmetry factors in the marcus normal and inverted regions. *Journal of Physical Chemistry B*, 107(19):4618–4623, 2003.

[14] H. Angerstein-Kozlowska, B. E. Conway, and W. B. A. Sharp. The real condition of electrochemical oxidized platinum surfaces. *Journal of Electroanalytical Chemistry*, 43:9–36, 1973.

[15] M. Aryanpour, A. Dhanda, and H. Pitsch. An algorithm for mass matrix calculation of constrained molecular geometries. *Journal of Chemical Physics*, 128(4), 2008.

[16] M. Aryanpour, V. Rai, and H. Pitsch. Convergent iterative constrained variation algorithm for calculation of electron-transfer transition states. *Journal of the Electrochemical Society*, 153(3):E52–7, 2006.

[17] Masoud Aryanpour. volmer05. *masoud.aryanpour(at)gmail.com*, 2006.

[18] A. J. Bard and L. R. Faulkner. *Electrochemical Methods: Fundamentals and Applications*. John Wiley & Sons, 2001.

[19] A.J. Bard. *Electrochemical Methods: Fundamentals and Applications*. New York, John Wiley, 2nd edition, 2001.

[20] J. Barrett. *Atomic Structure and Periodicity*. Wiley-Interscience, Cambridge, UK Royal Society of Chemistry, 2002.

[21] P. D. Beattie, V. I. Basura, and S. Holdcroft. Temperature and pressure dependence of o_2 reduction at pt/nafion 117 and pt/bam 407 interfaces. *Journal of Electroanalytical Chemistry*, 468:180–192, 1999.

[22] A. D. Becke. Density-functional exchange-energy approximation with correct asymptotic behavior. *Physical Review A*, 38(6):3098–100, 1988.

[23] M. Born and R. Oppenheimer. Zur quantentheorie der molekeln (on the quantum theory of molecules). *Annalen der Physik*, 84:457–484, 1927.

[24] G. B. Butler, K. F. O'Driscoll, and G. L. Wilkes. *JMS Macromolecular Chemistry and Physics*, C34:325–373, 1994.

[25] Y. Cai and A.B. Anderson. The reversible hydrogen electrode: Potential-dependent activation energies over platinum from quantum theory. *Journal of Physical Chemistry B*, 108(28):9829–9833, 2004.

[26] R.D. Cannon. *Electron Transfer Reactions*. London; Boston, Butterworths, 1980.

[27] L. Carrette, K.A. Friedrich, and U. Stimming. Improvement of CO tolerance of proton exchange membrane (PEM) fuel cells by a pulsing technique. *Fuel Cells*, 1(1), 2001.

[28] D. M. Ceperley and B. J. Alder. Ground state of the electron gas by a stochastic method. *Physical Review Letters*, 45(7):566–9, 1980.

[29] G. L. Closs, L. T. Calcaterra, N. J. Green, K. W. Penfield, and J. R. Miller. Distance, stereoelectronic effects, and the marcus inverted region in intramolecular electron-transfer in organic radical-anions. *Journal of Physical Chemistry*, 90(16):3673–3683, 1986.

[30] D.B. Cook. *Handbook of Computational Quantum Chemistry*. Dover Publications, Inc., Mineola, New York, 2005.

[31] Christopher J. Cramer. *Essentials of Computational Chemistry: Theories and Models*. John Wiley & Sons, Ltd, UK, second edition, 2004.

[32] A. Damjanovic and V. Brusic. Electrode kinetics of oxygen reduction on oxide free platinum electrodes. *Electrochimica Acta*, 12:615–628, 1967.

[33] A. Damjanovic, A. Dey, and J. Bockris. Kinetics of oxygen evolution and dissolution on platinum electrodes. *Electrochimica Acta*, 11:791–814, 1966.

[34] A. Damjanovic and P. G. Hudson. On the kinetics and mechanism of O_2 reduction at oxide film covered pt electrodes: Effect of oxide film thickness on kinetics. *Journal of Electrochemical Society*, 135(9):2269–2273, 1988.

[35] A. Eichler and J. Hafner. Reaction channels for the catalytic oxidation of CO on Pt(111). *Surface Science*, 435:58–62, 1999.

[36] A. Eisenberg, B. Hird, and R. B. Moore. New multiplet-cluster model for the morphology of random ionomers. *Macromolecules*, 23:4098, 1990.

[37] A Farazdel, M Dupuis, E Clementi, and A Aviram. Electric-field induced intermolecular electron-transfer in spiro π-electron systems and their suitability

as molecular electronic devices - a theoretical study. *Journal of the American Chemical Society*, 112(11):4206–4214, May 1990.

[38] P. J. Ferreira, G. J. la O, Y. Shao-Horn, D. Morgan, R. Makharia, S. Kocha, and H. A. Gasteiger. Instability of pt/c electrocatalysts in proton exchange membrane fuel cells: A mechanistic investigation. *Journal of Electrochemical Society*, 152(11):A2256–A2271, 2005.

[39] G.B. Fisher and Gland. The interaction of water with the Pt(111) surface. *Surface Science*, 94:446, 1980.

[40] GB Fisher and BA Sexton. Identification of an adsorbed hydroxyl species on the Pt(111) surface. *Physical Review Letters*, 44(10):683–6, 1980.

[41] M. Försth. Sensitivity analysis of the reaction mechanism for gas phase chemistry of H_2+O_2 mixture induced by a hot Pt surface. *Combustion and Flame*, 130:241–260, 2002.

[42] M. J. Frisch, G. W. Trucks, H. B. Schlegel, G. E. Scuseria, M. A. Robb, J. R. Cheeseman, J. A. Montgomery, Jr., T. Vreven, K. N. Kudin, J. C. Burant, J. M. Millam, S. S. Iyengar, J. Tomasi, V. Barone, B. Mennucci, M. Cossi, G. Scalmani, N. Rega, G. A. Petersson, H. Nakatsuji, M. Hada, M. Ehara, K. Toyota, R. Fukuda, J. Hasegawa, M. Ishida, T. Nakajima, Y. Honda, O. Kitao, H. Nakai, M. Klene, X. Li, J. E. Knox, H. P. Hratchian, J. B. Cross, V. Bakken, C. Adamo, J. Jaramillo, R. Gomperts, R. E. Stratmann, O. Yazyev, A. J. Austin, R. Cammi, C. Pomelli, J. W. Ochterski, P. Y. Ayala, K. Morokuma, G. A. Voth, P. Salvador, J. J. Dannenberg, V. G. Zakrzewski, S. Dapprich, A. D. Daniels, M. C. Strain, O. Farkas, D. K. Malick, A. D. Rabuck, K. Raghavachari, J. B. Foresman, J. V. Ortiz, Q. Cui, A. G. Baboul, S. Clifford, J. Cioslowski, B. B. Stefanov, G. Liu, A. Liashenko, P. Piskorz, I. Komaromi, R. L. Martin, D. J. Fox, T. Keith, M. A. Al-Laham, C. Y. Peng, A. Nanayakkara, M. Challacombe, P. M. W. Gill, B. Johnson, W. Chen, M. W. Wong, C. Gonzalez, and J. A. Pople. Gaussian 03, Revision C.02. Gaussian, Inc., Wallingford, CT, 2004.

[43] P. Fulde. *Electron Correlations in Molecules and Solids*. Springer-Verlag, Germany, 1991.

[44] T. D. Gierke, G. E. Munn, and F. C. Wilson. Morphology in nafion perfluorinated membrane products as determined by wide- and small-angle x-ray studies. *Journal of Polymer Science Part B-Polymer Physics*, 19:1687, 1981.

[45] H. Goldstein. *Classical Mechanics*. Addison-Wesley, 2nd edition, 1980.

[46] Z. H. Gu and P. B. Balbuena. Dissolution of oxygen reduction electrocatalysts in an acidic environment: Density functional theory study. *Journal of Physical Chemistry A*, 110(32):9783–9787, Aug 2006.

[47] V. Guerra and J. Loureiro. Dynamic Monte Carlo simulation of surface kinetics. *International Symposium on Rarefied Gas Dynamics*, 23:973–979, 2003.

[48] T. Hambleton, M. Aryanpour, and H. Pitsch. Analytical predictions of the changes in ionization potential due to a point charge. *internal report*, Summer 2005.

[49] P. J. Hay and W. R. Wadt. Ab initio effective core potentials for molecular calculations, potentials for the transition metal atoms Sc to Hg. *Journal of Chemical Physics*, 82(1):270–83, 1985.

[50] P. Hohenberg and W. Kohn. Inhomogeneous electron gas. *Physical Review*, 136(3B):B864–B871, 1964.

[51] W. Y. Hsu and T. D. Gierke. Elastic theory for ionic clustering in perfluorinated ionomers. *Macromolecules*, 15:101, 1982.

[52] T. Jacob. The mechanism of forming H_2O from H_2 and O_2 over a pt catalyst via direct oxygen reduction. *Fuel Cells*, 6(3/4):159 – 81, 2006.

[53] R. Kissel-Osterrieder, F. Behrendt, and J. Warnatz. Detailed modelling of the oxidation of CO on platinum: A Monte Carlo model. *Proc. Comb. Inst.*, 27:2267–2274, 1998.

[54] W. Koch and M.C. Holthausen. *A Chemist's Guide to Density Functional Theory*. WILEY-VCH,Germany, 2nd edition, 2000.

[55] L.N. Kostadinov and A.B. Anderson. Constrained variation calculations of electron-transfer transition states using the lagrange method. *Electrochemical and Solid-State Letters*, 6(10):E30–3, 2003.

[56] Chengteh Lee, Weitao Yang, and R. G. Parr. Development of the colle-salvetti correlation-energy formula into a functional of the electron density. *Physical Review B (Condensed Matter)*, 37(2):785–9, 1988.

[57] T. Li and P. B. Balbuena. Computational studies of the interactions of oxygen with platinum clusters. *Journal of Physical Chemistry B*, 105(41):9943–9952, Oct 2001.

[58] T. Li and P.B. Balbuena. Oxygen reduction on a platinum cluster. *Chemical Physics Letters*, 367(3/4):439–47, 2003.

[59] P. Liu and J. K. Norskov. Kinetics of the anode processes in PEM fuel cells: The promoting effect of Ru in PtRu anodes. *Fuel Cells*, 1(4):192–201, 2001.

[60] RA Marcus. On theory of electron-transfer reactions, unified treatment for homogenous and electrode reactions. *Journal of Chemical Physics*, 43(2):679, 1965.

[61] R.A. Marcus and N. Sutin. Electron transfers in chemistry and biology. *Biochimica et Biophysica Acta*, 811(3):265–322, 1985.

[62] N. M. Markovic, H. A. Gasteiger, and P. N. Ross. Oxygen reduction on platinum low-index single-crystal surfaces in sulfuric acid solution: Rotating Ring-Pt(hkl) Disk Studies. *Journal of Physical Chemistry*, 99(11):3411–3415, 1995.

[63] N. M. Markovic and P. N. Ross Jr. In A. Wieckowski, editor, *Interfacial Electrochemistry: Theory, Experiments and Applications*, page 821. Marcel Dekker, New York, 1999.

[64] N. M. Markovic, T. J. Schmidt, V. Stamenkovic, and P. N. Ross. *Fuel Cells*, 1(2):105–116, 2001.

[65] N.M. Markovic, T.J. Schmidt, V. Stamenkovic, and P.N. Ross. *Fuel Cells*, 1(2), 2001.

[66] K. A. Mauritz, C. J. Hora, and A. J. Hopfinger. Ions in polymers. In A. Eisenberg, editor, *ACS Advances in Chemistry*, volume 187, pages 124–154. American Chemical Society, 1980.

[67] D.A. McQuarrie and J.D. Simon. *Physical Chemistry: A Molecular Approach*. University Science Books, 1997.

[68] A. Michaelides and P. Hu. A density functional theory study of hydroxyl and the intermediate in the water formation reaction on pt. *Journal of Chemical Physics*, 114(1):513–19.

[69] A. Michaelides and P. Hu. Catalytic water formation on platinum: A first-principles study. *Journal of American Chemical Society*, 123(18):4235–4242, 2001.

[70] K. V. Mikkelsen and M. A. Ratner. Electron-tunneling in solid-state electron-transfer reactions. *Chemical Reviews*, 87(1):113–153, FEB 1987.

[71] R.J.D. Miller, G.L. McLendon, A.J. Nozik, W. Schmickler, and F. Willig. *Surface Electron Transfer Processes*. New York, VCH, 1995.

[72] M. D. Newton and N. Sutin. Electron-transfer reactions in condensed phases. *Annual Review of Physical Chemistry*, 35:437–480, 1984.

[73] MD Newton. Quantum chemical probes of electron-transfer kinetics - the nature of donor-acceptor interactions. *Chemical Reviews*, 91(5):767–792, 1991.

[74] K. C. Neyerlin, H. A. Gasteiger, C. K. Mittelsteadt, J. Jorne, and W. Gu. Effect of relative humidity on oxygen reduction kinetics in a pemfc. *Journal of Electrochemical Society*, 152(6):A1073–A1080, 2005.

[75] A. Nilekar, Y. Xu, J. Zhang, M. B. Vukmirovic, R. R. Adzic, and M. Mavrikakis. Experimental and theoretical studies of metal-supported pt monolayer catalysts for the oxygen reduction reaction. *AIChE Annual Meeting, Conference Proceedings*, pages 10619–10619, 2005 2005.

[76] J. K. Norskov, J. Rossmeisl, A. Logadottir, L. Lindqvist, J. R. Kitchin, T. Bligaard, and H. Jonsson. Origin of the overpotential for oxygen reduction at a fuel-cell cathode. *Journal of Physical Chemistry B*, 108(46):17886–17892, Nov 2004.

[77] R. O'Hyre, S.W. Cha, W. Colella, and F. Prinz. *Fuel Cell Fundamentals*. Wiley, 2005.

[78] Bockris J. O'M. and Khan U. M. *Surface Electrochemistry*. Plenum Press, New York, 1993.

[79] R.G. Parr and W. Yang. *Density-Functional Theory of Atoms and Molecules*. Oxford University Press, New York, 1989.

[80] A. Parthasarathy, B. Dave, S. Srinivasan, and J. Appleby. *Journal of Electrochemical Society*, 139(6):1634–1641, 1992.

[81] A. Parthasarathy, C. R. Martin, and S. Srinivasan. *Journal of Electrochemical Society*, 138:916–921, 1991.

[82] A. Parthasarathy, S. Srinivasan, and A. J. Appleby. Pressure dependence of the oxygen reduction reaction at the platinum microelectrode/nafion interface: Electrode kinetics and mass transport. *Journal of Electrochemical Society*, 139(10):2856–2862, 1992.

[83] A. Parthasarathy, S. Srinivasan, A. J. Appleby, and C. R. Martin. Temperature dependence of the electrode kinetics of oxygen reduction at the platinum/nafion interface - a microelectrode investigation. *Journal of Electrochemical Society*, 139(9):2530–2537, 1992.

[84] M. C. Payne, M. P. Teter, D. C. Allan, T. A. Arias, and J. D. Joannopoulos. Iterative minimization techniques for *ab-initio* total-energy calculation: Molecular dynamics and conjugate gradients. *Revies of Modern Physics*, 64(4):1045–1097, 1992.

[85] J. P. Perdew and M. R. Norman. Electron removal energies in kohn-sham density-functional theory. *Physical Review B (Condensed Matter)*, 26(10):5445–50, 1982.

[86] J. P. Perdew and W. Yue. Accurate and simple density functional for the electronic exchange energy: Generalized Gradient Approximation. *Physical Review B*, 33(12):8800–8802, 1986.

[87] H. Pitsch. Computational analyis of limitations of reactant consumptions in fuel cells. Annual report, Stanford University, 2005-2006.

[88] P. Politzer and F. Abu-Awwad. A comparative analysis of hartree-fock and kohn-sham orbital energies. *Theoretical Chemistry Accounts*, 99(2):83–7, 1998.

[89] P. Politzer, F. Abu-Awwad, and J. S. Murray. Comparison of density functional and hartree-fock average local ionization energies on molecular surfaces. *International Journal of Quantum Chemistry*, 69(4):607–13, 1998.

[90] S.R. Polo. Matrices D^{-1} and G^{-1} in the theory of molecular vibrations. *Journal of Chemical Physics*, 24(6):1133–1138, 1956.

[91] J. Rossmeisl, A. Logadottir, and J. K. Norskov. Electrolysis of water on (oxidized) metal surfaces. *Chemical Physics*, 319(1/3):178–84, DEC 2005.

[92] W. Schmickler. *Surface Electron Transfer Processes*, chapter Theory of Electrochemical Electron Transfer Reactions, pages 51–93. VCH Publishers, 1995.

[93] W. Schmickler. *Interfacial Electrochemistry*. New York, Oxford University Press, 1996.

[94] Y. Shao, G. Yin, Y. Gao, and P. Shi. Durability study of pt/c and pt/cnts catalysts under simulated pem fuel cell conditions. *Journal of Electrochemical Society*, 153(6):A1093–A1097, 2006.

[95] Reyimjan A. Sidik and Alfred B. Anderson. Density functional theory study of O_2 electroreduction when bonded to a pt dual site. *Journal of Electroanalytical Chemistry*, 528(1-2):69–76, Jun 2002.

[96] P. J. Stephens, F. J. Delvin, C. F. Chabalowski, and M. J. Frisch. Ab-initio calculation of vibrational absorption and circular-dichroism spectra using density functional force fields. *Journal of Physical Chemistry*, 98(45):11623–11627, 1994.

[97] P. Suppan. The Marcus Inverted Region. *Topics in Current Chemistry*, 163:95–130, 1992.

[98] A. Szabo and N.S. Ostland. *Modern Quantum Chemistry, Introduction to Advanced Electronic Structure Theory*. Dover Publications, Inc., Mineola, New York, 1996.

[99] Jr. T. H. Dunning. Gaussian basis sets for use in correlated molecular calculations. i. the atoms boron through neon and hydrogen. *Journal of Chemical Physics*, 90(2):1007–23, 1989.

[100] C. D. Taylor and M. Neurock. Theoretical insights into the structure and reactivity of the aqueous/metal interface. *Current Opinion in Solid State and Materials Science*, 9(1-2):49–65, February/April 2005.

[101] M. Aryanpour V. Rai and H. Pitsch. A first-principles approach for quantification of potential-dependent adsorbate interactions: Application to water discharge mechanism on platinum(111). *212th ECS Meeting - Washington, DC*, 524, October 2007.

[102] B. Viswanathan and M. Aulice Scibioh. *Fuel Cells: Principles And Applications*. Universities Press, India.

[103] S.H. Vosko, L. Wilk, and M. Nusair. Accurate spin-dependent electron liquid correlation energies for local spin density calculations: a critical analysis. *Canadian Journal of Physics*, 58(8):1200–11, 1980.

[104] W. R. Wadt and P. J. Hay. Ab initio effective core potentials for molecular calculations, potentials for main group elements Na to B. *Journal of Chemical Physics*, 82(1):284–98, 1985.

[105] F. T. Wagner and P. N. Ross Jr. Leed analysis of electrode surfaces: Structural effects of potentiodynamic cycling on pt single crystals. *Journal of Electroanalytical Chemistry*, 150:141–164, 1983.

[106] N. Wakabayashi, M. Takeichi, M. Itagaki, H. Uchida, and M. Watanabe. Temperature-dependence of oxygen reduction activity at a platinum electrode in an acidic electrolyte solution investigated with a channel flow double electrode. *Journal of Electroanalytical Chemistry*, 574(2):339–346, 2005.

[107] S. Walch, A. Dhanda, M. Aryanpour, and H. Pitsch. Mechanism of molecular oxygen reduction at the cathode of a PEM fuel cell: Non-electrochemical reactions on catalytic pt particles. *Physical Chemistry, (in press)*, 2008.

[108] J. X. Wang, N. M. Markovic, and R. R. Adzic. Kinetic analysis of oxygen reduction on Pt(111) in acid solutions: Intrinsic kinetic parameters and anion adsorption *Journal of Physical Chemistry B*, 108:4127–4133, 2004.

[109] Y. Wang and P. Balbuena. Roles of proton and electric field in the electroreduction of O_2 on Pt(111) surfaces: Results of an ab-initio molecular dynamics study. *Journal of Physical Chemistry B*, 108(14):4376–4384, 2004.

[110] E.B. Wilson, J.C. Decius, and P.C. Cross. *Molecular Vibrations: The Theory of Infrared and Raman Vibrational Spectra*. Dover, New York, 1980.

[111] H. S. Wroblowa, M. L. B. Rao, A. Damjanovic, and J. O'M. Bockris. *Journal of Electroanalytical Chemistry*, 1:139–150, 1967.

[112] H. Xu, Y. Song, H. R. Kunz, and J. M. Fenton. Effect of elevated temperature for reduced relative humidity on orr kinetics for pem fuel cells. *Journal of Electrochemical Society*, 152(9):A1828–A1836, 2005.

[113] K. Yasuda, A. Taniguchi, T. Akita, T. Ioroi, and Z. Siroma. Platinum dissolution and deposition in the polymer electrolyte membrane of a pem fuel cell as studied by potential cycling. *Chemical Physics*, 8:746–752, 2006.

[114] E. Yeager. Electrocatalysts for O_2 reduction. *Electrochimica Acta*, 29:1527, 1984.

[115] H. L. Yeager and A. J. Steck. Cation and water diffusion in nafion ion exchange membranes: Influence of polymer structure. *Journal of Electrochemical Society*, 128:1880–1884, 1981.

[116] V. P. Zhdanov. Electrochemical reactions on catalyst particles with three-phase boundaries. *Physical Review E*, 67:042601–1–042601–4, 2003.

[117] V. P. Zhdanov and B. Kasemo. Towards the understanding of the specifics of reactions in polymer-electrolyte fuel cells. *Surface Science*, 554(2/3):103–8, Apr 2004.

[118] V. P. Zhdanov and B. Kasemo. Simulation of the charge and potential distribution in the double layer formed by polymer electrolyte. *Electrochemistry Communications*, 8(4):561–4, April 2006.

[119] C. F. Zinola, A. M. Castro Luna, and A. J. Arvia. Temperature dependence of kinetic parameters related to oxygen electroreduction in acid solutins on platinum electrodes. *Electrochimica Acta*, 39(13):1951–1959, 1994.

www.ingramcontent.com/pod-product-compliance
Lightning Source LLC
LaVergne TN
LVHW022307060326
832902LV00020B/3329